# Lecture Notes in Mathematics

**Editors:**
J.-M. Morel, Cachan
F. Takens, Groningen
B. Teissier, Paris

Krzysztof Bogdan · Tomasz Byczkowski
Tadeusz Kulczycki · Michal Ryznar
Renming Song · Zoran Vondraček

# Potential Analysis of Stable Processes and its Extensions

Volume Editors:
Piotr Graczyk
Andrzej Stos

 Springer

*Editors*
Piotr Graczyk
LAREMA
Université d'Angers
2 bd Lavoisier
49045 Angers
France
Piotr.Graczyk@univ-angers.fr

Andrzej Stos
Laboratoire de Mathématiques
Université Blaise Pascal
Campus Universitaire des Cézeaux
63177 Aubière
France
stos@math.univ-bpclermont.fr

*Authors: see List of Contributors*

ISBN: 978-3-642-02140-4        e-ISBN: 978-3-642-02141-1
DOI: 10.1007/978-3-642-02141-1

Lecture Notes in Mathematics ISSN print edition: 0075-8434
ISSN electronic edition: 1617-9692

Library of Congress Control Number: 2009928106

Mathematics Subject Classification (2000): 60J45, 60G52, 60J50, 60J75, 31B25, 31C05, 31C35, 31C25

*Cover design*: SPi Publisher Services

Printed on acid-free paper

springer.com

# Foreword

This monograph is devoted to the potential theory of stable stochastic processes and related topics, such as the subordinate Brownian motions (including the relativistic process) and Feynman–Kac semigroups generated by certain Schrödinger operators.

The stable Lévy processes and related stochastic processes play an important role in stochastic modelling in applied sciences, in particular in financial mathematics, and the theoretical motivation for the study of their fine properties is also very strong. The potential theory of stable and related processes naturally extends the theory established in the classical case of the Brownian motion and the Laplace operator.

The foundations and general setting of probabilistic potential theory were given by G.A. Hunt [92](1957), R.M. Blumenthal and R.K. Getoor [23](1968), S.C. Port and J.C. Stone [130](1971). K.L. Chung and Z. Zhao [62](1995) have studied the potential theory of the Brownian motion and related Schrödinger operators. The present book focuses on classes of processes that contain the Brownian motion as a special case. A part of this volume may also be viewed as a probabilistic counterpart of the book of N.S. Landkof [117](1972).

The main part of Introduction that opens the book is a general presentation of fundamental objects of the potential theory of the isotropic stable Lévy processes in comparison with those of the Brownian motion (presented in a subsection). The introduction is accessible to a non-specialist. Also the chapters that follow should be of interest to a wider audience. A detailed description of the content of the book is given at the end of Chapter 1.

Some of the material of the book was presented by T. Byczkowski, T. Kulczycki, M. Ryznar and Z. Vondraček at the Workshop on Stochastic and Harmonic Analysis of Processes with Jumps held at Angers, France, May 2-9, 2006. The authors are grateful to the organizers and to the main supporters of the Workshop – the CNRS, the European Network of Harmonic Analysis HARP and the University of Angers – for this opportunity, which gave the incentive to write the monograph.

The book was written while Z. Vondraček was visiting the Department of Mathematics of University of Illinois at Urbana-Champaign. He thanks the department for the stimulating environment and hospitality. Thanks are also due to Andreas Kyprianou for several useful comments. The editors thank T. Luks for critical reading of some parts of the manuscript and for some of the figures illustrating the text.

# Contents

# List of Contributors

**Krzysztof Bogdan** Institute of Mathematics and Computer Science Wrocław University of Technology, ul. Wybrzeże Wyspiańskiego 27, 50-370 Wrocław, Poland. Krzysztof.Bogdan@pwr.wroc.pl

The research of this author was partially supported by grant MNiI 1 P03A 026 29.

**Tomasz Byczkowski** Institute of Mathematics and Computer Science, Wrocław University of Technology, ul. Wybrzeże Wyspiańskiego 27, 50-370 Wrocław, Poland. Tomasz.Byczkowski@pwr.wroc.pl

The research of this author was partially supported by KBN Grant 1 P03A 020 28 and RTN Harmonic Analysis and Related Problems, contract HPRN-CT-2001-00273-HARP.

**Tadeusz Kulczycki** Institute of Mathematics, Polish Academy of Sciences, ul. Kopernika 18, 51-617 Wrocław, Poland and Institute of Mathematics and Computer Science, Wrocław University of Technology, ul. Wybrzeże Wyspiańskiego 27, 50-370 Wrocław, Poland. TKulczycki@impan.pan.wroc.pl

The research of this author was partially supported by KBN Grant 1 P03A 020 28 and RTN Harmonic Analysis and Related Problems, contract HPRN-CT-2001-00273-HARP.

**Michał Ryznar** Institute of Mathematics and Computer Science, Wrocław University of Technology, ul. Wybrzeże Wyspiańskiego 27, 50-370 Wrocław, Poland. Michal.Ryznar@pwr.wroc.pl

The research of this author was partially supported by KBN Grant 1 P03A 020 28 and RTN Harmonic Analysis and Related Problems, contract HPRN-CT-2001-00273-HARP.

**Renming Song** Department of Mathematics, University of Illinois, Urbana, IL 61801. rsong@math.uiuc.edu

The research of this author is supported in part by a joint US-Croatia grant INT 0302167.

**Zoran Vondraček** Department of Mathematics, University of Zagreb, Zagreb, Croatia. vondra@math.hr

The research of this author is supported in part by MZOS grant 037-0372790-2801 of the Republic of Croatia.

# Chapter 1
# Introduction

## 1.1 Bases of Potential Theory of Stable Processes

In 1957, G. A. Hunt introduced and developed the potential theory of Markov processes in his fundamental treatise [92]. Hunt's theory is essentially based on the fact that the integral of the transition probability of a Markov process defines a **potential kernel**:

$$U(x,y) = \int_0^\infty p(t,x,y)dt\,.$$

One of the important topics in the theory is the study of multiplicative functionals of the Markov process, corresponding either to **Schrödinger perturbations** of the generator of the process, or to **killing the process** at certain stopping times. Among the most influential treatises on this subject are the monographs [23] by R. M. Blumenthal and R. K. Getoor, [60] by K. L. Chung, [22] by W. Hansen and J. Bliedtner, and [62] by K. L. Chung and Z. Zhao.

**Harmonic functions** of a strong Markov process are defined by the mean value property with respect to the distribution of the process stopped at the first exit time of a domain. An important case of such a function is the potential of a measure not charging the domain, thus yielding no "sources" to change the expected occupation time of the process.

To produce specific results, however, the general framework of Hunt's theory requires precise information on the asymptotics of the potential kernel of the given Markov process. For instance, the process of the Brownian motion in $\mathbb{R}^3$ is generated by the Laplacian, $\Delta$, and yields the Newtonian kernel, $x \mapsto c|x-y|^{-1}$. Here $y$ is the source or pole of the kernel. When $x_0$ is fixed and $|y| \to \infty$, we have that, regardless of $x$, $|x-y|^{-1}/|x_0-y|^{-1} \to 1$, which eventually leads to the conclusion that nonnegative functions harmonic on the whole of $\mathbb{R}^3$ must be constant.

Explicit formulas for the potential kernel are rare. Even the Brownian motion killed when first exiting a subdomain of $\mathbb{R}^d$ in general leads to a

K. Bogdan et al., *Potential Analysis of Stable Processes and its Extensions*,
Lecture Notes in Mathematics 1980, DOI 10.1007/978-3-642-02141-1_1,
© Springer-Verlag Berlin Heidelberg 2009

transition density and potential kernel which are not given by closed-form formulas, and may be even difficult to estimate.

A primary example of a jump process is the isotropic $\alpha$-stable Lévy process in $\mathbb{R}^d$, whose potential kernel is the M. Riesz' kernel. The analytic theory of the Riesz kernel, the fractional Laplacian $\Delta^{\alpha/2}$, and the corresponding $\alpha$-harmonic functions had been well established for a long time (see [133] and [117]). However, until recently little was known about the boundary behavior of $\alpha$-harmonic functions on sub-domains of $\mathbb{R}^d$.

We begin the book by presenting some of the basic objects and results of the classical (Newtonian) potential theory ($\alpha = 2$), and Riesz potential theory ($0 < \alpha < 2$). We have already mentioned the well known but remarkable fact that the (Newtonian) potential theory of the Laplacian can be interpreted and developed by means of the Brownian motion ([71]). An analogous relationship holds for the (Riesz) potential theory of the fractional Laplacian and the isotropic $\alpha$-stable Lévy process. We pursue this relationship in the following sections. We like to remark that $\Delta^{\alpha/2}$ is a primary example of a nonlocal pseudo differential operator ([97]) and we hope that a part of our discussion will extend to other nonlocal operators. Apart from its significance in mathematics, the fractional Laplacian appears in theoretical physics in the connection to the problem of stability of the matter [118]. Namely, the operator $I - (I - \Delta)^{1/2}$ corresponds to the kinetic energy of a relativistic particle and $\Delta^{1/2}$ can be regarded as an approximation to $I - (I - \Delta)^{1/2}$, see, e.g., [45], [134].

In what follows, functions and sets are assumed to be Borel measurable. We will write $f \approx g$ to indicate that $f$ and $g$ are *comparable*, i.e. there is a constant $c$ (a positive real number independent of $x$), such that $c^{-1}f(x) \leqslant g(x) \leqslant cf(x)$. Values of *constants* may change from place to place, for instance $f(x) \leqslant (2c+1)g(x) = cg(x)$ should not alarm the reader.

### 1.1.1  Classical Potential Theory

We consider the Gaussian kernel,

$$g_t(x) = \frac{1}{(4\pi t)^{d/2}}\, e^{-|x|^2/4t}\,, \quad x \in \mathbb{R}^d\,, \quad t > 0\,. \tag{1.1}$$

It is well known that $\{g_t,\, t \geqslant 0\}$ form a convolution semigroup: $g_s * g_t = g_{s+t}$, where $s, t > 0$. This property is at the heart of the classical potential theory. Complicating the notation slightly, we define transition probability

$$g(s, x, t, A) = \int_{A-x} g_{t-s}(y)dy\,, \quad s < t\,, \ x \in \mathbb{R}^d\,, \ A \subset \mathbb{R}^d\,. \tag{1.2}$$

The semigroup property of $\{g_t\}$ is equivalent to the following Chapman-Kolmogorov equation

$$\int_{\mathbb{R}^d} g(s,x,u,dz)g(u,z,t,A) = g(s,x,t,A)\,, \quad s < u < t\,, \ x \in \mathbb{R}^d\,, \ A \subset \mathbb{R}^d\,.$$

If $d \geqslant 3$ then we define and calculate the Newtonian kernel,

$$N(x) = \int_0^\infty g_t(x)dt = \mathcal{A}_{d,2}|x|^{2-d}\,, \quad x \in \mathbb{R}^d\,.$$

Here and below

$$\mathcal{A}_{d,\gamma} = \Gamma((d-\gamma)/2)/(2^\gamma \pi^{d/2}|\Gamma(\gamma/2)|)\,. \tag{1.3}$$

The semigroup property yields that $N * g_s(x) = \int_s^\infty g_t(x)dt \leqslant N(x)$. Recall that a function $h \in C^2(D)$ is called *harmonic* in an open set $D \subseteq \mathbb{R}^d$ if it satisfies Laplace's equation,

$$\Delta h(x) = \sum_{i=1}^d \frac{\partial^2 h(x)}{\partial x_i^2} = 0\,, \quad x \in D\,. \tag{1.4}$$

It is well known that $N$ is harmonic on $\mathbb{R}^d \setminus \{0\}$. Let $B(a,r) = \{x \in \mathbb{R}^d : |x-a| < r\}$, where $a \in \mathbb{R}^d$, $r > 0$. We also let $B_r = B(0,r)$, $B = B_1 = B(0,1)$. The *Poisson kernel* of $B(a,r)$ is

$$P(x,z) = \frac{\Gamma(d/2)}{2\pi^{d/2}r} \frac{r^2 - |x-a|^2}{|x-z|^d}\,, \quad x \in B(a,r)\,, \quad z \in \partial B(a,r)\,. \tag{1.5}$$

It is well known that if $h$ is harmonic in an open set containing the closure of $B(a,r)$ then

$$h(x) = \int_{\partial B(a,r)} h(z)\,P(x,z)\sigma(dz)\,, \quad x \in B(a,r)\,. \tag{1.6}$$

Here $\sigma$ denotes the $(d-1)$-dimensional Haussdorff measure on $\partial B(a,r)$. We like to note that $P(x,z)$ is positive and continuous on $B(a,r) \times \partial B(a,r)$, and has the following properties:

$$\int_{\partial B(a,r)} P(x,z)\sigma(dz) = 1\,, \quad x \in B(a,r)\,, \tag{1.7}$$

$$\lim_{x \to w} \int_{\partial B(a,r) \setminus B(w,\delta)} P(x,z)\sigma(dz) = 0\,, \quad w \in \partial B(a,r)\,, \quad \delta > 0\,. \tag{1.8}$$

It is also well known that for every $z \in \partial B(a,r)$, $P(\cdot, z)$ is harmonic in $B(a,r)$, a property resembling Chapman-Kolmogorov equation if we consider (1.6) for $h(x) = P(x, z_0)$. Consequently, if $f \in C(\partial B(a,r))$, then the Poisson integral,

$$P[f](x) = \int_{\partial B(a,r)} P(x,z) f(z)\sigma(dz), \quad x \in B(a,r), \qquad (1.9)$$

solves the *Dirichlet problem* for $B(a,r)$ and $f$. Namely, $P[f]$ extends to the unique continuous function on $B(a,r) \cup \partial B(a,r)$, which is harmonic in $B(a,r)$, and coincides with $f$ on $\partial B(a,r)$, see (1.8). In particular, $P[1] \equiv 1$, compare (1.7).

An analogous *Martin representation* is valid for every nonnegative $h$ harmonic on $B(a,r)$,

$$h(x) = P[\mu](x) := \int_{\partial B(a,r)} P(x,z)\,\mu(dz), \quad x \in B(a,r). \qquad (1.10)$$

Here $\mu \geqslant 0$ is a unique nonnegative measure on $\partial B(a,r)$. We like to note that appropriate *sections* of $P[\mu]$ *weakly* converge to $\mu$ ([107]), which reminds us that in general the boundary values of harmonic functions require handling with care.

By (1.5) we have that $P(x_1, z) \leqslant (1 + s/r)^d (1 - s/r)^{-d} P(x_2, z)$ if $x_1, x_2 \in B(a,s)$, $s < r$, $z \in \partial B(a,r)$. As a direct application of (1.10) we obtain the following *Harnack inequality*,

$$c^{-1}h(x_1) \leqslant h(x_2) \leqslant c\, h(x_1), \quad x_1, x_2 \in B(a,s), \qquad (1.11)$$

provided $h$ is *nonnegative* harmonic. We see that $h$ is *nearly constant* (i.e. comparable with 1) on $B(a,s)$ for $s < r$. If $D$ is *connected*, then considering finite coverings of *compact* $K \subset D$ by *overlapping chains* of balls, we see that nonnegative functions $h$ harmonic on $D$ are nearly constant on $K$, see Figure 1.1.

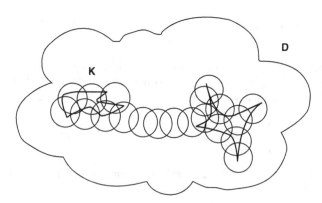

**Fig. 1.1** Harnack chain

Despite its general importance, Harnack inequality is less useful at the boundary of the domain because the corresponding constant gets inflated for points close to the boundary. In fact, nonnegative harmonic functions present a complicated array of asymptotic behaviors at the boundary, see (1.5). To study the asymptotics, we first concentrate on nonnegative harmonic function *vanishing* at a part of the boundary.

The Boundary Harnack Principle (**BHP**) for *classical* harmonic functions delicately depends on the geometric regularity of the domain. To simplify our discussion we will consider the following Lipschitz condition. Let $d \geqslant 2$. Recall that $\Gamma : \mathbb{R}^{d-1} \to \mathbb{R}$ is called Lipschitz if there is $\lambda < \infty$ such that

$$|\Gamma(y) - \Gamma(z)| \leqslant \lambda |z - y|, \quad y, z \in \mathbb{R}^{d-1}. \tag{1.12}$$

We define (special Lipschitz domain)

$$D_\Gamma = \{x = (x_1, \ldots, x_d) \in \mathbb{R}^d : x_d > \Gamma(x_1, \ldots, x_{d-1})\}. \tag{1.13}$$

A nonempty open $D \subseteq \mathbb{R}^d$ is called a Lipschitz domain if for every $z \in \partial D$ there exist $r > 0$, a Lipschitz function $\Gamma : \mathbb{R}^{d-1} \to \mathbb{R}$, and an isometry $T$ of $\mathbb{R}^d$, such that $D \cap B(z, r) = T(D_\Gamma) \cap B(z, r)$, that is, if $D$ is locally isometric with a set "above" the graph of a Lipschitz function.

**Theorem 1.1 (Boundary Harnack Principle).** *Let $D$ be a connected Lipschitz domain. Let $U \subset \mathbb{R}^d$ be open and let $K \subset U$ be compact. There exists $C < \infty$ such that for every (nonzero) functions $u, v \geqslant 0$, which are harmonic in $D$ and vanish continuously on $D^c \cap U$, we have*

$$C^{-1} \frac{u(y)}{v(y)} \leqslant \frac{u(x)}{v(x)} \leqslant C \frac{u(y)}{v(y)}, \quad x, y \in K \cap D. \tag{1.14}$$

Thus, the ratio $u/v$ is *nearly constant* on $D \cap K$. Furthermore, under the above assumptions,

$$\lim_{x \to z} \frac{u(x)}{v(x)} \text{ exists as } x \to z \in \partial D \cap K, \tag{1.15}$$

see Figure 1.2.

The theorem is crucial in the study of asymptotics and structure of general nonnegative harmonic functions in Lipschitz domains. The proof of **BHP** for classical harmonic functions in Lipschitz domains was independently given by B. Dahlberg(1977), A. Ancona(1978) and J.-M. Wu(1978), and (1.15) was published by D. Jerison and C. Kenig in 1982.

We now return to $\{g_t\}$, and the resulting transition probability $g$. By Wiener's theorem there are probability measures $P^x$, $x \in \mathbb{R}^d$, on the space of all *continuous* functions (paths) $[0, \infty) \ni t \mapsto X(t) \in \mathbb{R}^d$, such that $P^{x_0}(X(0) = x_0) = 1$ and $P^{x_0}(X_t \in A | X_s = x) = g(s, x, t, A) = P^x(X_{t-s} \in$

**Fig. 1.2** The setup of
**BHP**

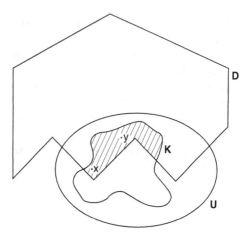

A), for $x_0, x \in \mathbb{R}^d$, $0 \leqslant s < t$, $A \subset \mathbb{R}^d$. Recall that the construction of the distribution of the process from a transition probability on this path space requires certain continuity properties of the transition probability in time. Here we have $\lim_{t \to 0} g_t(x)/t = 0$ $(x \neq 0)$, which eventually allows the paths to be continuous by Kolmogorov's test or by Kinney-Dynkin theorem, see [135]. Thus, $X_t$ is continuous. Denote $X = (X_t) = (X_t^1, \ldots, X_t^d)$, and let $E^x$ be the integration with respect to $P^x$. We have $E^0 X_t^i = 0$, $E^0 (X_t^i)^2 = 2t$. Thus, $X_t = B_{2t}$, where $B_t$ is the usual Brownian motion with variance of each coordinate equal to $t$.

By the construction, $E^x f(X_t) = \int_{\mathbb{R}^d} f(y) g(0, x, t, dy) = \int_{\mathbb{R}^d} f(y) g_t(y - x) dy$ for $x \in \mathbb{R}^d$, $t > 0$, and nonnegative or integrable $f$. For a (Borel) set $A \subset \mathbb{R}^d$ by Fubini-Tonelli theorem,

$$E^x \int_0^\infty \mathbf{1}_A(X_t) dt = \int_A N(y - x) dy, \quad x \in \mathbb{R}^d.$$

Therefore $N(\cdot - x)$ may be interpreted as the density function of (the measure of) the expected occupation time of the process, when started at $x$.

So far we have only considered $X$ evaluated at constant (deterministic) times $t$. For an open $D \subseteq \mathbb{R}^d$ we now define the *first exit time* from $D$,

$$\tau_D = \inf\{t > 0 : X_t \notin D\}.$$

By the usual convention, $\inf \emptyset = \infty$. $\tau_D$ is a Markov (stopping) time. A function $h$ defined and Borel measurable on $\mathbb{R}^d$ is harmonic on $D$ if for every open bounded set $U$ such that $\overline{U} \subseteq D$ (denoted $U \subset\subset D$) we have

$$h(x) = E^x h(X_{\tau_U}), \quad x \in U. \tag{1.16}$$

We assume here the absolute convergence of the integral. Since the $P^x$-distribution of $X_{\tau_{B(x,r)}}$ is the normalized surface measure on the sphere $\partial B(x, r)$, the equality (1.16) reads as follows:

$$h(x) = \int_{\partial B(x,r)} h(y)\, P(0, y - x)\sigma(dy)\,,$$

if $x \in U = B(a, r)$, see (1.5) and (1.6). Thus, (1.16) agrees with the classical definition of harmonicity.

The above definitions may and will be extended below to other strong Markov processes, and (1.16) may be referred to as the "averaging property" or "mean value property".

We should note that (for the Brownian motion) the values of $h$ on $D^c$ are irrelevant in (1.16) because $X_{\tau_U} \in \partial U \subset D$ in (1.16). For the isotropic stable Lévy process, which we will discuss below, the support of the distribution of the process stopped at the first exit time of a domain is typically *the whole complement* of the domain. Indeed, as time ($t$) advances, the paths of the process may leave the domain either by *continuously approaching* the boundary or by a direct *jump* to the complement of the domain. In particular, a harmonic function should generally be defined on the whole of $\mathbb{R}^d$. It is of considerable importance to classify nonnegative harmonic functions of the process according to these two scenarios, see the concluding remarks in [38].

To indicate the role of the strong Markov property, we consider a nonnegative function $\tilde{h}$ on $D^c$ and we let $h(x) = E^x \tilde{h}(X_{\tau_D})$, $x \in D$. We will regard $\tilde{h}$ on $D^c$ as the boundary/external values of $h$, as appropriate for general processes with jumps. It will be convenient to write $h(x)$ for $\tilde{h}(x)$ if $x \in D^c$. Let $x \in U \subset\subset D$. We have

$$E^x h(X_{\tau_U}) = E^x E^{X_{\tau_U}} h(X_{\tau_D}) = E^x h(X_{\tau_D}) = h(x)\,.$$

In particular we see that $h$ is harmonic on $U$. The above essentially also proves that $\{h(X_{\tau_U})\}$ is a martingale ordered by the inclusion of (open relatively compact) subsets $U$ of $D$, with respect to every $P^x$, $x \in D$. Closability of such martingales is of some interest in this theory [27, 38], and relates to the existence of boundary values of harmonic functions. For instance the martingales given by Poisson integrals (1.10) are not closable for singular measures $\mu$ on $\partial B(a, r)$.

### 1.1.2 Potential Theory of the Riesz Kernel

We will introduce the principal object of this book, namely the isotropic (rotation invariant) $\alpha$-stable Lévy process. We will construct the transition density of the process by using convolution semigroups of measures. For a measure $\gamma$ on $\mathbb{R}^d$, we let $|\gamma|$ denote its total mass. For a function $f$ we let $\gamma(f) = \int f d\gamma$, whenever the integral makes sense. When $|\gamma| < \infty$ and $n = 1, 2, \ldots$ we let $\gamma^n = \gamma * \ldots * \gamma$ ($n$ times) denote the $n$-fold convolution of $\gamma$ with itself:

$$\gamma^n(f) = \int f(x_1 + x_2 + \cdots + x_n)\gamma(dx_1)\gamma(dx_2)\ldots\gamma(dx_n).$$

We also let $\gamma^0 = \delta_0$, the evaluation at 0. If $\gamma$ is finite on $\mathbb{R}^d$ then we define

$$P_t^\gamma = \exp t(\gamma - |\gamma|\delta_0) := \sum_{n=0}^\infty \frac{t^n (\gamma - |\gamma|\delta_0)^n}{n!} \tag{1.17}$$

$$= (\exp -t|\gamma|\delta_0) * \exp t\gamma = e^{-t|\gamma|} \sum_{n=0}^\infty \frac{t^n \gamma^n}{n!}, \quad t \in \mathbb{R}. \tag{1.18}$$

By (1.18) each $P_t^\gamma$ is a probability measure, provided $\gamma \geqslant 0$ and $t \geqslant 0$, which we will assume in what follows. By (1.17), $P_t^\gamma$ form a convolution semigroup,

$$P_t^\gamma * P_s^\gamma = P_{s+t}^\gamma, \quad s, t \geqslant 0.$$

Furthermore, for two such measures $\gamma_1$, $\gamma_2$, we have

$$P_t^{\gamma_1} * P_t^{\gamma_2} = P_t^{\gamma_1 + \gamma_2}, \quad t > 0.$$

By (1.17),

$$\lim_{t \to 0}(P_t^\gamma - \delta_0)/t = \gamma - |\gamma|\delta_0. \tag{1.19}$$

In the following discussion for simplicity we will also assume that $\gamma$ has bounded support and that $\gamma$ is symmetric: $\gamma(-A) = \gamma(A)$, $A \subset \mathbb{R}^d$. The reader may want to verify that

$$\int_{\mathbb{R}^d} |y|^2 P_t^\gamma(dy) = t \int_{\mathbb{R}^d} |y|^2 \gamma(dy) < \infty, \quad t \geqslant 0. \tag{1.20}$$

As a hint we note that only the third term in (1.17) contributes to (1.20). In particular,

$$P_t^\gamma(B(0, R)^c) \leqslant t \int_{\mathbb{R}^d} |y|^2 \gamma(dy)/R^2 \to 0 \quad \text{as } R \to \infty. \tag{1.21}$$

We define

$$\nu(B) = \mathcal{A}_{d,-\alpha} \int_B |z|^{-d-\alpha} dz, \quad B \subset \mathbb{R}^d. \tag{1.22}$$

It is a Lévy measure, i.e. a nonnegative measure on $\mathbb{R}^d \setminus \{0\}$ satisfying

$$\int_{\mathbb{R}^d} \min(|y|^2, 1)\, \nu(dy) < \infty. \tag{1.23}$$

We also note that $\nu$ is symmetric. We consider the following operator, the fractional Laplacian,

$$\Delta^{\alpha/2} u(x) = \mathcal{A}_{d,-\alpha} \lim_{\varepsilon \to 0^+} \int_{\{y \in \mathbb{R}^d:\, |y-x|>\varepsilon\}} \frac{u(y) - u(x)}{|y - x|^{d+\alpha}}\, dy. \tag{1.24}$$

The limit exists if, say, $u$ is $C^2$ near $x$ and bounded on $\mathbb{R}^d$. The claim follows from Taylor expansion of $u$ at $x$, with remainder of order two, and by the symmetry of $\nu$. We like to note that $A = \Delta^{\alpha/2}$ satisfies the *positive maximum principle*: for every $\varphi \in C_c^\infty(\mathbb{R}^d)$

$$\sup_{y \in \mathbb{R}^d} \varphi(y) = \varphi(x) \geqslant 0 \quad implies \quad A\varphi(x) \leqslant 0.$$

The most general operators on $C_c^\infty(\mathbb{R}^d)$ which have this property are of the form

$$A\varphi(x) = \sum_{i,j=1}^d a_{ij}(x) D_{x_i} D_{x_j} \varphi(x) + b(x)\nabla\varphi(x) + q(x)\varphi(x)$$
$$+ \int_{\mathbb{R}^d} \left( \varphi(x+y) - \varphi(x) - y\nabla\varphi(x)\, \mathbf{1}_{|y|<1} \right) \mu(x, dy). \tag{1.25}$$

Here $y\nabla\varphi$ is the scalar product of $y$ and the gradient of $\varphi$, and for every $x$, $a(x) = (a_{ij}(x))_{i,j=1}^n$ is a real nonnegative definite symmetric matrix, the vector $b(x) = (b_i(x))_{i=1}^d$ has real coordinates, $q(x) \leqslant 0$, and $\mu(x, \cdot)$ is a Lévy measure. The description is due to Courrège, see [90, Proposition 2.10], [151, Chapter 2] or [97, Chapter 4.5]. For translation invariant operators of this type, $a$, $b$, $q$, and $\mu$ are independent of $x$. For $\Delta^{\alpha/2}$ we further have $a = 0$, $b = 0$, $q = 0$ and $\mu = \nu$.

For $r > 0$ and a function $\varphi$ on $\mathbb{R}^d$ we consider its dilation $\varphi_r(y) = \varphi(y/r)$, and we note that $\nu(\varphi_r) = r^{-\alpha}\nu(\varphi)$. In particular, $\nu$ is homogeneous: $\nu(rB) = r^{-\alpha}\nu(B)$, $B \subset \mathbb{R}^d$. Similarly, if $\varphi \in C_c^\infty(\mathbb{R}^d)$, then $\Delta^{\alpha/2}(\varphi_r) = r^{-\alpha}(\Delta^{\alpha/2}\varphi)_r$.

We will consider approximations of $\nu$ and $\Delta^{\alpha/2}$ suggested by (1.24). For $0 < \delta \leqslant \varepsilon \leqslant \infty$ we define measures $\nu_{\delta,\varepsilon}(f) = \int_{\delta \leqslant |y| < \varepsilon} f(y)\nu(dy)$. We have

$$P_t^{\nu_{\delta,\infty}} - P_t^{\nu_{\varepsilon,\infty}} = P_t^{\nu_{\varepsilon,\infty}} * \left( P_t^{\nu_{\delta,\varepsilon}} - \delta_0 \right). \tag{1.26}$$

When $\varepsilon \to 0$, the above converges (uniformly in $\delta$) to 0 on each $C_c^\infty$ function with compact support. This claim follows from Taylor expansion with the quadratic remainder, (1.20) applied to $\gamma = \nu_{\delta,\varepsilon}$, (1.23), and the fact that $P_t^{\nu_\varepsilon,\infty}$ are probabilities, hence uniformly finite.

If $\phi$ is a bounded continuous function on $\mathbb{R}^d$, $\eta > 0$, and $0 < R < \infty$, then there is $\varphi \in C_c^\infty$ such that $|\phi - \varphi| < \eta$ on $B(0,R)$. We have

$$
\begin{aligned}
\left| \left( P_t^{\nu_\delta,\infty} - P_t^{\nu_\varepsilon,\infty} \right)(\phi) \right| \leqslant \; & \left| \left( P_t^{\nu_\delta,\infty} - P_t^{\nu_\varepsilon,\infty} \right)(\varphi) \right| + 2\eta \\
& + \left[ P_t^{\nu_1,\infty} * P_t^{\nu_\delta,1}(B(0,R)^c) \right. \\
& \left. + P_t^{\nu_1,\infty} * P_t^{\nu_\varepsilon,1}(B(0,R)^c) \right] \sup |\phi - \varphi| \, .
\end{aligned}
$$

By inspecting (1.21) we see that the measures $P_t^{\nu_\varepsilon,\infty}$ weakly converge to a probability measure, say $P_t$, as $\varepsilon \to 0$, and so $\{P_t, t \geqslant 0\}$ is a convolution semigroup, too. We also note that $P_t/t$ weakly converges to $\nu$ on (closed subsets of) $\mathbb{R}^d \setminus \{0\}$. This follows from the approximation of $P_t$ by $P_t^{\nu_\varepsilon,\infty}$, and (1.19).

The Fourier transform of $P_t^\varepsilon$ is easily calculated from (1.17),

$$
\widehat{P_t^{\nu_\varepsilon,\infty}}(u) = \int_{\mathbb{R}^d} e^{iuy} P_t^{\nu_\varepsilon,\infty}(dy) = \exp\left( t \int_{\mathbb{R}^d} (e^{iuy} - 1)\nu_{\varepsilon,\infty}(dy) \right), \quad u \in \mathbb{R}^d,
\tag{1.27}
$$

hence $\widehat{P_t}(u) = \exp(t\Phi(u))$, where

$$
\begin{aligned}
\Phi(u) &= \int_{\mathbb{R}^d} \left( e^{iuy} - 1 - iuy \mathbf{1}_{B(0,1)}(y) \right) \nu(dy) \\
&= \int_{\mathbb{R}^d} (\cos(uy) - 1)\, \nu(dy) = -\frac{\pi}{2 \sin \frac{\pi\alpha}{2} \Gamma(1+\alpha)} \int_{|\xi|=1} |u\xi|^\alpha \sigma(d\xi) = -c|u|^\alpha.
\end{aligned}
\tag{1.28}
$$

In fact, $c = 1$ here, and $\Phi(u) = -|u|^\alpha$ as we shall see momentarily.

For $t > 0$, the measures $P_t$ have rapidly decreasing Fourier transform hence they are absolutely continuous with bounded smooth densities, $p_t(x)$, given by the Fourier inversion formula:

$$
p_t(x) = (2\pi)^{-d} \int_{\mathbb{R}^d} e^{ixu}\, e^{-t|u|^\alpha}\, du\,.
\tag{1.29}
$$

Explicit formulas for the $p_t$ do not exist except for $\alpha = 1$,

$$
p_t(x) = \frac{\Gamma\left(\frac{d+1}{2}\right)}{\pi^{\frac{d+1}{2}}} \frac{t}{(t^2 + |x|^2)^{\frac{d+1}{2}}},
$$

and $\alpha = 2$, which corresponds to the Brownian motion and is excluded from our present considerations. Clearly, $p_s * p_t(x) = p_{s+t}(x)$, $s, t > 0$. From (1.28) we obtain the *scaling property*:

$$
p_t(x) = t^{-d/\alpha} p_1(t^{-1/\alpha} x)\,, \quad x \in \mathbb{R}^d,\ t > 0.
\tag{1.30}
$$

In particular,

$$p_t(x) \leqslant ct^{-d/\alpha}. \tag{1.31}$$

We define the *potential kernel* (M. Riesz kernel) of the convolution semigroup of functions $\{p_t\}$:

$$K_\alpha(x) = \int_0^\infty p_t(x)dt, \quad x \in \mathbb{R}^d. \tag{1.32}$$

When $d > \alpha$, we have that $K_\alpha(x)$ is finite for $x \neq 0$, see (1.31). By (1.30),

$$K_\alpha(x) = \mathcal{A}_{d,\alpha}|x|^{\alpha-d}. \tag{1.33}$$

The explicit constant here and in (1.28) may be obtained by a calculation involving Bessel functions, but it is easier to employ the Laplace transform to this end (see below). Since $K_\alpha \equiv \infty$ if $\alpha \geqslant d$, and cumbersome modifications are needed in this case, the dimension $d = 1$ is explicitly excluded from our considerations. We refer to [41] for more information and references on the case $d = 1$. *Unless stated otherwise in the remainder of this chapter we assume that $d \geqslant 2$.*

We construct the *standard* isotropic $\alpha$-stable Lévy process $(Y_t, P^x)$ in $\mathbb{R}^d$ by specifying the following density function of its transition probability:

$$p(s, x, t, y) := p_{t-s}(y - x), \quad x, y \in \mathbb{R}^d, \quad s < t,$$

and stipulating that $P^x(Y(0) = x) = 1$. This is completely analogous to the construction of the Brownian motion, except for the fact that the distribution of the process is eventually concentrated on right continuous paths with left limits (rather than on continuous paths). The latter follows from the Kinney-Dynkin theorem and the fact that $P_t/t$ converges to $\nu \neq 0$. In fact $Y_t$ has discontinuities of the first kind, that is jumps (by $z$), occurring in time ($t$) with intensity $\nu(dz)dt$ ([135]), see Figure 1.3. The term *standard* above involves certain technical measure-theoretic and topological assumptions, involving the right continuity and left-limitedness of the paths $t \mapsto Y_t$ as mentioned above, and the so called quasi-left continuity, see, e.g., [18], [23] for more details. The process $(Y_t, P^x)$ is Markov on $\mathbb{R}^d$ with transition probabilities $P^{x_0}(Y_t \in A | Y_s = x) = \int_A p(s, x, t, y)dy$, for $x_0, x \in \mathbb{R}^d$, $0 \leqslant s < t$, $A \subset \mathbb{R}^d$; and initial distribution is specified by $P^{x_0}(Y(0) = x_0) = 1$. It is also well-known that $(Y_t, P^x)$ is strong Markov with respect to the so-called standard filtration ([23]). As usual, $P^x$, $E^x$ denote the distribution and expectation for the process starting from $x$. We note that by the symmetry of ($\nu$ and) $P_t$, $E^x f(Y_t) = E^0 f(x + Y_t) = \int_{\mathbb{R}^d} f(x + y)P_t(dy) = f * p_t(x)$ and

$$\widehat{(f * p_t)}(\xi) = \hat{f}(\xi)e^{-t|\xi|^\alpha}. \tag{1.34}$$

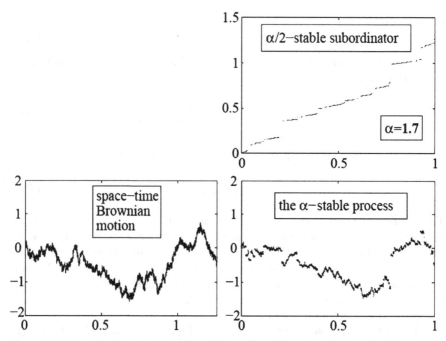

**Fig. 1.3** Subordination: trajectories of the $\alpha/2$-stable subordinator, the Brownian motion, and the $\alpha$-stable Lévy process in $\mathbb{R}$

The (semigroup of the) process $Y_t$ has $\Delta^{\alpha/2}$ as the infinitesimal generator ([156, 97]). Indeed, by using (1.19) and (1.26) we have that

$$\Delta^{\alpha/2}u(x) = \lim_{t \downarrow 0} \frac{E^x u(Y_t) - u(x)}{t} \tag{1.35}$$

$$= \mathcal{A}_{d,-\alpha} \lim_{\varepsilon \to 0^+} \int_{\{y \in \mathbb{R}^d \colon |y-x| > \varepsilon\}} \frac{u(y) - u(x)}{|y-x|^{d+\alpha}} dy, \quad x \in \mathbb{R}^d, \, u \in C_c^\infty(\mathbb{R}^d).$$

The result can also be obtained by using Fourier inversion formula and (1.34) ([90, 97]). To justify the notation $\Delta^{\alpha/2}$, we note that the Fourier symbol of $\Delta^{\alpha/2}$, and the Fourier symbol of the Laplacian regarded as convolution-type (i.e. translation invariant) operators satisfy the equation

$$-\widehat{\Delta^{\alpha/2}}(\xi) = |\xi|^\alpha = \left(\widehat{(-\Delta)}(\xi)\right)^{\alpha/2}, \tag{1.36}$$

compare (1.35), (1.34).

We will briefly recall an alternative method of constructing $p_t$ (and $X_t$) by subordination of the Gaussian kernel (and the Brownian motion, respectively). For $\beta \in (0, 1)$ we denote by $\{\eta_t^\beta\}$ the standard $\beta$-stable subordinator, i.e. the nondecreasing Lévy process on the line starting at 0, and determined by the Laplace transform

$$\mathbf{E}e^{-u\eta_t^\beta} = e^{-tu^\beta}, \quad t \geqslant 0, \quad u > 0. \tag{1.37}$$

Here $\mathbf{E}$ is the expectation corresponding to $\eta^\beta$. Let $h_\beta(t, y)$ denote the transition density of the process, for $t > 0$. The function can be obtained either from Haussdorff-Bernstein-Widder theorem on completely monotone functions and (1.37), or from an explicit construction based on (1.17), which gives (1.37) from the following Lévy measure on $\mathbb{R}$:

$$1_{\{y>0\}} \frac{\beta}{\Gamma(1-\beta)} y^{-1-\beta} dy,$$

see also [86] for precise asymptotics of the function. The potential kernel of the subordinator is

$$\int_0^\infty h_\beta(t, y)\, dt = \frac{y^{\beta-1}}{\Gamma(\beta)}, \quad y > 0, \tag{1.38}$$

which is verified by applying the Laplace transform to each side of (1.38), and by using (1.37). By (1.38) and Fubini's theorem,

$$\mathbf{E} \int_0^\infty f(\eta_t^\beta)\, dt = \frac{1}{\Gamma(\beta)} \int_0^\infty y^{\beta-1} f(y)\, dy. \tag{1.39}$$

We consider $\beta = \alpha/2 \in (0, 1)$. Let $\{X_t\}$ be the Brownian motion and assume that $X$ and $\eta^{\alpha/2}$ are stochastically independent.

We may now define

$$p_t(x) = \int_0^\infty g_u(x) h_{\alpha/2}(t, u)\, du,$$

and

$$Y_t = X_{\eta_t^{\alpha/2}}.$$

Let $P^x$, $E^x$ denote the resulting (i.e. product) probability measure, and the corresponding expectation. Here the Brownian motion (hence also $Y$) starts from $x \in \mathbb{R}^d$. We will write $E = E^0$ and we denote by $\tilde{E}$ the expectation for the Brownian motion starting at 0. In view of the independence of $X$ and $\eta^{\alpha/2}$ we have

$$Ee^{iY_t\xi} = \mathbf{E}\tilde{E}\exp\left[iX_{\eta_t^{\alpha/2}}\xi\right] = \mathbf{E}e^{-\eta_t^{\alpha/2}|\xi|^2} = e^{-t(|\xi|^2)^{\alpha/2}} = e^{-t|\xi|^\alpha}, \tag{1.40}$$

compare (1.28). We can also identify the constant in the definition of the Riesz kernel (1.33), that is prove that for $\alpha < d$ the potential operator of $Y$

has $K_\alpha$ as the (convolutional) kernel,

$$U_\alpha f(x) = E^x \int_0^\infty f(Y_t)\, dt = \mathcal{A}_{d,\alpha} \int_{\mathbb{R}^d} |y - x|^{\alpha - d} f(y)\, dy\,. \qquad (1.41)$$

Indeed, by subordination and (1.38) we have

$$
\begin{aligned}
E^x \int_0^\infty f(X_{\eta_t^{\alpha/2}})\, dt &= \int_0^\infty Ef(x + X_{\eta_t^{\alpha/2}})\, dt \\
&= \int_{\mathbb{R}^d} \int_0^\infty \int_0^\infty f(x + y)\, g_u(y)\, h_{\alpha/2}(t, u)\, du\, dy\, dt \\
&= \int_{\mathbb{R}^d} f(x+y) \int_0^\infty \left\{ \int_0^\infty h_{\alpha/2}(t, u)\, dt \right\} (4\pi u)^{-d/2} e^{-|y|^2/4u}\, du\, dy \\
&= \frac{1}{2^d \pi^{d/2} \Gamma(\alpha/2)} \int_{\mathbb{R}^d} f(z) \int_0^\infty u^{\alpha/2 - d/2 - 1} e^{-|z - x|^2/4u}\, du\, dz \\
&= \mathcal{A}_{d,\alpha} \int_{\mathbb{R}^d} \frac{f(z)}{|z - x|^{d-\alpha}}\, dz = f * K_\alpha(x)\,,
\end{aligned}
$$

compare (1.3). The Fourier symbol of the operator of convolution with $K_\alpha$ is $|\xi|^{-\alpha}$.

In order to effectively study the potential theory of $Y$ on domains $D \subset \mathbb{R}^d$, we need explicit formulas, or at least estimates for the *potential kernel* of the process killed at the first instant of leaving $D$, that is for the *Green function* for the fractional Laplacian on $D$.

### 1.1.3  Green Function and Poisson Kernel of $\Delta^{\alpha/2}$

The Green function and the harmonic measure of the fractional Laplacian are defined in [117, Theorem IV.4.16, pp. 229, 240], see also [25], [22, pp. 191, 250, 384], [109], and [129], [41] for the case of dimension one. We will briefly recall the construction. The finite Green function $G_D(x, y)$ of $D$, if it exists (e.g., if $d > \alpha$ or $D$ is bounded), is bound to satisfy

$$\int_{\mathbb{R}^d} G_D(x, v) \Delta^{\alpha/2} \varphi(v)\, dv = -\varphi(x)\,, \quad x \in \mathbb{R}^d\,, \ \varphi \in C_c^\infty(D)\,. \qquad (1.42)$$

For instance, if $d > \alpha$ then (1.42) holds for $D = \mathbb{R}^d$ and the Riesz kernel:

$$G_{\mathbb{R}^d}(x, y) = K_\alpha(y - x)\,, \quad x, y \in \mathbb{R}^d\,, \qquad (1.43)$$

as follows from the inspection of their Fourier symbols, see [117, (1.1.12')], or from the approximation (1.17). Thus, $G_D$ is the integral kernel of the inverse of $-\Delta^{\alpha/2}$ on $C_c^\infty(D)$. We like to remark that within the framework of the theory of fractional powers of *nonnegative* operators, the preferred notation for $\Delta^{\alpha/2}$ is $-(-\Delta)^{\alpha/2}$ (and so $G_D = \left(-(-(-\Delta)^{\alpha/2})\right)^{-1}$). In the following discussion we will stick to our previous (shorter) notation, and we will keep assuming that $d \geqslant 2$, in particular $d > \alpha$. Within this setup, $\omega_D^x$, the harmonic measure of $D$, is defined as the unique ([117, p. 245], [41]) subprobability measure (probability measure if $D$ is bounded) concentrated on $D^c$ such that $\int_{\mathbb{R}^d} G_{\mathbb{R}^d}(z,y)\omega_D^x(dz) \leqslant G_{\mathbb{R}^d}(x,y)$ for all $y \in \mathbb{R}^d$, and

$$\int_{\mathbb{R}^d} G_{\mathbb{R}^d}(z,y)\omega_D^x(dz) = G_{\mathbb{R}^d}(x,y) \qquad (1.44)$$

.for $y \in D^c$ (except at points of $\partial D$ *irregular* for the Dirichlet problem on $D$, see [117], [23]).

Given $\omega_D$, the Green function can be defined *pointwise* as

$$G_D(x,y) = G_{\mathbb{R}^d}(x,y) - \int_{D^c} G_{\mathbb{R}^d}(z,y)\omega_D^x(dz). \qquad (1.45)$$

More generally, we have

$$G_D(x,v) = G_U(x,v) + \int_{\mathbb{R}^d} G_D(w,v)\omega_U^x(dw), \qquad x,v \in \mathbb{R}^d, \text{ if } U \subset D. \quad (1.46)$$

As seen in [38], (1.46) is the origin of the notion of harmonicity. In particular, $x \mapsto G_D(x,y)$ is $\alpha$-harmonic on $D \setminus \{y\}$. The symmetry of $G_{\mathbb{R}^d}(x,y) = K_\alpha(y-x) = G_{\mathbb{R}^d}(y,x)$ implies that $G_D(x,v) = G_D(v,x)$ for $x,v \in \mathbb{R}^d$ ([117, p. 285]), which is related to Hunt's switching equality [23], and eventually dates back to George Green's work on potential of electric fields. We recall that the harmonic measure is the $P^x$-distribution of $Y_{\tau_D}$, the process stopped at the first instance of leaving $D$,

$$\int f(y)\omega^x(dy) = E^x f(Y_{\tau_D}), \qquad x \in \mathbb{R}^d,$$

and the Green function is the density function of the integral kernel of the **Green operator**,

$$E^x \int_0^{\tau_D} f(Y_t)dt = \int_{\mathbb{R}^d} f(y)G_D(x,y)dy =: G_D f(x), \qquad x \in \mathbb{R}^d.$$

Indeed, these statements result from the following application of the strong Markov property,

$$G_{\mathbb{R}^d} f(x) = E^x \int_0^\infty f(Y_t)dt = E^x \int_0^{\tau_D} f(Y_t)dt$$

$$+ E^x \int_{\tau_D}^\infty f(Y_t)dt = G_D f(x) + E^x G_{\mathbb{R}^d} f(Y_{\tau_D}).$$

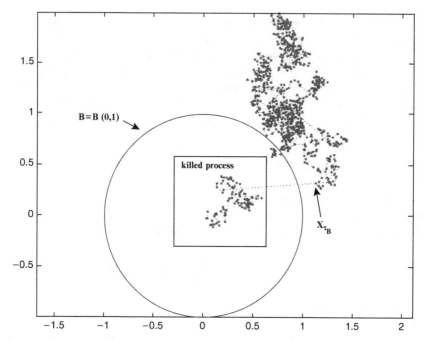

**Fig. 1.4** Trajectory of the stable process leaving the unit disc on the plane, $\alpha = 1.8$

see (1.41, 1.43, 1.44, 1.45) and Figure 1.4.

We can interpret $y \mapsto G_D(x, y)$ as the density function of the expected time spent at $y$ by the process $Y$ started at $x$, before it exits $D$ for the first time.

Using (1.42), (1.46), and Fubini's theorem we obtain

$$\int_D G_D(x, v)\Delta^{\alpha/2}\varphi(v)dv = \int_{D^c} [\varphi(y) - \varphi(x)]\omega_D^x(dy), \quad x \in \mathbb{R}^d, \, \varphi \in C_c^\infty(\mathbb{R}^d).$$
(1.47)

We define the expected first exit time of $D$ (by $Y$),

$$s_D(x) = E^x \tau_D = \int_{\mathbb{R}^d} G_D(x, v)dv, \quad x \in \mathbb{R}^d,$$
(1.48)

and the Poisson kernel of $D$:

$$P_D(x, y) = \int_D G_D(x, v)\mathcal{A}_{d, -\alpha}|y - v|^{-d-\alpha} \, dv, \quad x \in \mathbb{R}^d, \, y \in D^c. \quad (1.49)$$

Consider (open) $U \subset D$. By integrating (1.46) against the Lebesgue measure we obtain

$$s_D(x) = s_U(x) + \int_{\mathbb{R}^d} s_D(y) \omega_U^x(dy), \quad x \in \mathbb{R}^d, \; U \subset D. \tag{1.50}$$

Integrating (1.46) against $\mathcal{A}_{d,-\alpha} |y - v|^{-d-\alpha} \, dv$ on $\mathbb{R}^d$, we get

$$P_D(x,y) = P_U(x,y) + \int P_D(z,y) \omega_U^x(dz), \quad x \in U, \; y \in D^c. \tag{1.51}$$

The reader may attempt interpreting (1.51) as harmonicity of $P_D(x,y) + \delta_y$, see [38]. By considering $\varphi$ approximating $\mathbf{1}_A$ for *open* $A \subset D^c$ in (1.47), and by (1.24) and Fubini's theorem we arrive at the Ikeda-Watanabe formula ([95]):

$$\omega_D^x(A) = \int_A P_D(x,y) dy, \quad x \in D. \tag{1.52}$$

We can interpret (1.49), and (1.52), as follows. Jumping from $v \in D$ to $y \in D^c$ happens over time with intensity $\mathcal{A}_{d,-\alpha} |y - v|^{-d-\alpha}$. The intensity is integrated against the occupation time measure, $G_D(x,v)dv$, thus giving $P_D(x,y)$.

For a large class of domains (but not for all domains) $\omega_D^x(\partial D) = 0$ for every $x \in D$, so that

$$\omega_D^x(dy) = P_D(x,y)dy \;\; \text{on } D^c. \tag{1.53}$$

In particular, (1.53) holds for domains with the outer cone property (a class of open sets containing finite intersections of bounded Lipschitz domains). This is proved by noting that $P^x(Y_{\tau_{B_x}} \in D^c)$ is bounded away from 0 for $x \in D$ and $B_x = B(x, \frac{1}{2}\text{dist}(x, D^c))$, see the formula (1.57) below. Thus the process $Y$ started at $x$ will leave $D$ by a jump (from within $B_x$) with a positive probability. If $Y_{\tau_{B_x}}$ jumps to $D \setminus B_x$, then this reasoning can be repeated by the strong Markov property. Since leaving $D$ *continuously* requires an infinite number of such jumps (from balls $B_x$ to $D \setminus B_x$), the probability of continuous approach to the boundary is zero, see [27], [152], [154]. Below we will use the observation for the intersection of a given Lipschitz domain with a ball, see Figure 2.1.

We note in passing that (1.47) and (1.53) yield the following decomposition of each $\varphi \in C_c^\infty(\mathbb{R}^d)$ into a sum of a Green potential and Poisson integral,

$$\varphi(x) = \int_D G_D(x,v) \left[ -\Delta^{\alpha/2} \varphi(v) \right] dv + \int_{D^c} P_D(x,y) \varphi(y) dy, \quad x \in D, \tag{1.54}$$

where we assumed that $D$ is bounded, hence $\omega_D^x(\mathbb{R}^d) = 1$. One can interpret the two integrals as resulting from "sources" within $D$, and *jumps* between

$D$ and $D^c$ ("tunneling" from sources outside of $D$). For general domains $D$ the picture is somewhat complicated by an additional term related to the continuous approach of $Y_t$ to $\partial D$ at $t = \tau_D$, see [38] and [125] for details.

The Green function of the ball is known explicitly:

$$G_{B_r}(x, v) = \mathcal{B}_{d,\alpha} |x - v|^{\alpha - d} \int_0^w \frac{s^{\alpha/2 - 1}}{(s + 1)^{d/2}} \, ds \,, \quad x, v \in B_r \,, \qquad (1.55)$$

where

$$w = (r^2 - |x|^2)(r^2 - |v|^2)/|x - v|^2$$

and $\mathcal{B}_{d,\alpha} = \Gamma(d/2)/(2^\alpha \pi^{d/2} [\Gamma(\alpha/2)]^2)$, see [25], [133].

Also ([78], [33], [43]),

$$s_{B_r}(x) = \frac{C_\alpha^d}{\mathcal{A}_{d,-\alpha}} (r^2 - |x|^2)^{\alpha/2} \,, \quad |x| \leqslant r \,, \qquad (1.56)$$

and

$$P_{B_r}(x, y) = C_\alpha^d \left( \frac{r^2 - |x|^2}{|y|^2 - r^2} \right)^{\alpha/2} \frac{1}{|x - y|^d} \,, \quad |x| < r, \; |y| > r \,, \qquad (1.57)$$

where $C_\alpha^d = \pi^{1+d/2} \Gamma(d/2) \sin(\pi \alpha/2)$, see Figures 1.5 and 1.6.

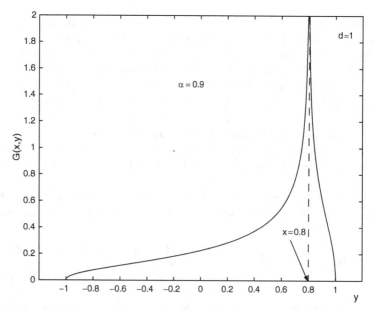

**Fig. 1.5** Green function of $(-1, 1)$

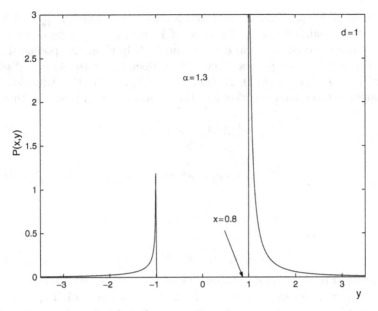

**Fig. 1.6** Poisson kernel of $(-1, 1)$

The formulas (1.55) and (1.57) are essentially due to Marcel Riesz and date back to 1938 ([133]). They were completed and interpreted in the present framework by R. Blumenthal, R. Getoor and D. Ray in 1961 ([25]). The proof of (1.55) and (1.57) consists of verification of (1.44) for $\omega_{B_r}^x(dy) = 1_{B_r^c}(y)P_r(x, y)dy$. This is a rather involved procedure, aided by the use of the *inversion*, $Tx = |x|^{-2}x$, where $x \in \mathbb{R}^d \setminus \{0\}$. We refer to [25], [117] and [133] for details of the calculation. The explanation of the role of the inversion (and the corresponding Kelvin transform) in obtaining the Poisson kernel of the ball, along with the interpretation of the inversion in terms of the process $Y$ are given in [41]. For instance ([41]),

$$G_{TD}(x, y) = |x|^{\alpha-d}|y|^{\alpha-d}G_D(Tx, Ty). \qquad (1.58)$$

Here $TD = \{Tx : x \in D\}$. We encourage the reader to verify (1.58) for $D = \mathbb{R}^d$.

Results similar to (1.55) and (1.57) exist for the complement of the ball and for half-spaces. In fact they can be obtained from (1.55) and (1.57) by using (1.58). For half-spaces a different proof of (1.55) and (1.57) (i.e. one not using the Kelvin transform) was obtained recently as a consequence of the corresponding results for the relativistic process in [44]. The formula (1.56) was first given in [78], see also [33]. For clarity, we need to emphasize that the distribution of $Y_{\tau_B(x,r)}$, given by (1.57), is concentrated on $\{y : |y - x| > r\}$, hence with probability one the first exit of $Y$ from $B(x,r)$ is *by a jump*.

Apart from inversion, *scaling* is extremely helpful in the potential theory of the stable Lévy process. Let $r > 0$. Recall that $\nu(rA) = r^{-\alpha}\nu(A)$, $\Delta^{\alpha/2}\varphi_r(x) = r^{-\alpha}\Delta^{\alpha/2}\varphi(x/r)$, $K_\alpha(rx) = r^{\alpha-d}K_\alpha(x)$. By the definition and uniqueness of the Green function and the harmonic measure we see that

$$\omega_{rD}^{rx}(rA) = \omega_D^x(A)\,, \tag{1.59}$$

and

$$G_{rD}(rx, rv) = r^{\alpha-d}G_D(x,v)\,, \tag{1.60}$$

hence

$$s_{rD}(rx) = r^\alpha s_D(x)\,, \tag{1.61}$$

and

$$P_{rD}(rx, ry) = r^{-d}P_D(x,y)\,. \tag{1.62}$$

*Translation invariance* and *rotation invariance* are equally important but easier to observe. For example $G_{D+y}(x + y, v + y) = G_D(x, v)$. Together with scaling and inversion the properties help reduce many of our considerations to the setting of the unit ball centered at the origin.

A function $u$ defined (and Borel measurable) on $\mathbb{R}^d$ is called $\alpha$-**harmonic** in an (open) set $D$ if it is harmonic on $D$ for the isotropic $\alpha$-stable Lévy process $Y$: for every open bounded set $U$ such that $\overline{U} \subseteq D$ we have

$$u(x) = E^x u(Y_{\tau_U})\,, \quad x \in U\,. \tag{1.63}$$

We assume here absolute convergence of the integral.

A counterpart of Weyl's lemma holds ([32]) for $\alpha$-harmonic functions: $u$ is $\alpha$-harmonic in $D$ if and *only if* $u$ is $C^2$ on $D$, and

$$\Delta^{\alpha/2}u(x) = 0\,, \quad x \in D\,. \tag{1.64}$$

We note that for this condition to hold, $u$ must be defined on the whole of $\mathbb{R}^d$, and its values on $D^c$ are crucial for this property. This reflects the fact that $\Delta^{\alpha/2}$ is a non-local operator allowing for a direct influence between distant points $x$ and $y$ in the domain of $u$, see (1.24). In particular, the following integrability assumption holds for $\alpha$-harmonic function $u$:

$$\int_{\mathbb{R}^d} \frac{|u(y)|}{(1 + |y|)^{d+\alpha}}\,dy < \infty\,.$$

We may also define $\Delta^{\alpha/2}$ in a weak (distributional) sense. This allows to consider $\Delta^{\alpha/2}$ and (Schrödinger operators) $\Delta^{\alpha/2}\phi + q\phi$ as defined locally, i.e. on arbitrary open sets (in the sense of the L. Schwartz' theory of distributions), even for discontinuous $q$ which do not have well-defined pointwise values (see below). In this connection we like to mention the following observation: if $u$ is $\alpha$-harmonic on open $U_1$ and $u$ is $\alpha$-harmonic on open $U_2$ then $u$ is $\alpha$-harmonic on $U_1 \cup U_2$. This fact is trivial according to (1.64). It can also be proved by using the probabilistic definition of $\alpha$-harmonicity (1.63), but such proof is no longer trivial ([33]). This points out a local aspect of the otherwise *non-local* property of $\alpha$-harmonicity (the reader interested in more details may consult [32] and [33]). In the remainder of this survey we will not employ the *weak* fractional Laplacian (we refer the interested reader to [32] for details of this approach to Schrödinger operators). We will use a probabilistic methodology based on the so-called multiplicative functionals and Green functions, rather than on infinitesimal generators; and we refer the interested reader to [33] for more.

If $u$ is $\alpha$-harmonic in a domain containing $\overline{B(0,r)}$ then

$$u(x) = C_\alpha^d \int_{|y|>r} \left[ \frac{r^2 - |x|^2}{|y|^2 - r^2} \right]^{\alpha/2} |y - x|^{-d} u(y) dy, \quad x \in B(0,r). \quad (1.65)$$

We see that such $u$ is $C^\infty$ on $B(0,r)$. For nonnegative $u$ we also obtain **Harnack inequality**. The next two propositions are versions of it.

**Proposition 1.2.** *If $u \geqslant 0$ on $\mathbb{R}^d$ and $u$ is $\alpha$-harmonic on $D \supset B_\rho \supset B_r \ni x_1, x_2$, then*

$$u(x_1) \leqslant \left( \frac{1 + r/\rho}{1 - r/\rho} \right)^d u(x_2). \quad (1.66)$$

*Proof.* If $r \leqslant s < \rho$ then by (1.57) we have $P_{B_s}(x_1, z) \leqslant (1 + r/s)^d (1 - r/s)^{-d} P_{B_s}(x_2, z)$ for $|z| \geqslant s$. Using (1.65) (for $B(0,s)$)), and letting $s \to \rho$, we prove the result. □

**Proposition 1.3.** *If $x_1, x_2 \in D$ then there is $c_{x_1,x_2}$ such that for every $\lambda \geqslant 0$*

$$u(x_1) \leqslant c_{x_1,x_2} u(x_2). \quad (1.67)$$

*Proof.* If $x_1, x_2 \in B_r \subset B_{2r} \subset D$ for some $r > 0$ then we are done by Lemma 1.2 with $c = c_{x_1,x_2}$ depending only on $d$. Assume that $B(x_1, 2r) \subset D$, $B(x_2, 2r) \subset D$, $B(x_1, 2r) \cap B(x_2, 2r) = \emptyset$ for some $r > 0$, and consider (1.63) with $U = B(x_1, r)$. Let $y \in D^c$. By (1.65) and the first part of the proof we obtain $u(x_1) \geqslant c \int_{B(x_2,r)} u(x_2) P_{B_r}(0, x - x_1) dx = cu(x_2)$. □

If $K \subset D$ is compact and $x_1, x_2 \in K$ then $c_{x_1,x_2}$ in Harnack's inequality above may be so chosen to depend only on $K$, $D$, and $\alpha$, because $r$ in the above proof may be chosen independently of $x_1, x_2$. Note that $D$ and $K$ may be *disconnected*. This shows a certain advantage of the fact that $Y$ has jumps.

## 1.1.4   Subordinate Brownian Motions

The rotationally invariant $\alpha$–stable processes are obtained from the Brownian motion by a subordination procedure.

Let $X = (X(u) : u \geq 0)$ be a $d$-dimensional Brownian motion. Subordination of Brownian motion consists of time-changing the paths of $X$ by an independent subordinator. To be more precise, let $S = (S_t : t \geq 0)$ be a subordinator (i.e., a nonnegative, increasing Lévy process) independent of $X$. The process $Y = (Y_t : t \geq 0)$ defined by $Y_t = X(S_t)$ is called a subordinate Brownian motion. The process $Y$ is an example of a rotationally invariant $d$-dimensional Lévy process. A general Lévy process in $\mathbb{R}^d$ is completely characterized by its characteristic triple $(b, A, \pi)$, where $b \in \mathbb{R}^d$, $A$ is a nonnegative definite $d \times d$ matrix, and $\pi$ is a measure on $\mathbb{R}^d \setminus \{0\}$ satisfying $\int (1 \wedge |x|^2) \, \pi(dx) < \infty$, called the Lévy measure of the process. Its characteristic exponent $\Phi$, defined by $\mathbb{E}[\exp\{i\langle x, Y_t\rangle\}] = \exp\{-t\Phi(x)\}$, $x \in \mathbb{R}^d$, is given by the Lévy-Khintchine formula involving the characteristic triple $(b, A, \pi)$. The main difficulty in studying general Lévy processes stems from the fact that the Lévy measure $\pi$ can be quite complicated.

The situation simplifies immensely in the case of subordinate Brownian motions. If we take the Brownian motion $X$ as given, then $Y$ is completely determined by the subordinator $S$. Hence, one can deduce properties of $Y$ from properties of the subordinator $S$. On the analytic level this translates to the following: Let $\phi$ denote the Laplace exponent of the subordinator $S$. That is, $\mathbb{E}[\exp\{-\lambda S_t\}] = \exp\{-t\phi(\lambda)\}$, $\lambda > 0$. Then the characteristic exponent $\Phi$ of the subordinate Brownian motion $Y$ takes on the very simple form $\Phi(x) = \phi(|x|^2)$ (our Brownian motion $X$ runs at twice the usual speed). Hence, properties of $Y$ should follow from properties of the Laplace exponent $\phi$. This will be one of main themes of these lecture notes – we will study potential-theoretic properties of $Y$ by using information given by $\phi$. Results obtained by this approach include explicit formulas for the Green function of $Y$ and the Lévy measure of $Y$. Let $p(t, x, y)$, $x, y \in \mathbb{R}^d$, $t > 0$, denote the transition densities of the Brownian motion $X$, and let $\mu$, respectively $U$, denote the Lévy measure, respectively the potential measure, of the subordinator $S$. Then the Lévy measure $\pi$ of $Y$ is given by $\pi(dx) = J(x) \, dx$, where

$$J(x) = \int_0^\infty p(t, 0, x) \, \mu(dt),$$

while the Green function $G(x, y)$, $x, y \in \mathbb{R}^d$, of $Y$ is given by

$$G(x, y) = \int_0^\infty p(t, x, y) \, U(dt).$$

Let us consider the second formula (same reasoning also applies to the first one). This formula suggests that the asymptotic behavior of $G(x, y)$ when

$|x-y| \to 0$ (respectively, when $|x-y| \to \infty$) should follow from the asymptotic behavior of the potential measure $U$ at $\infty$ (respectively at 0). The latter can be studied in the case when the potential measure has a monotone density $u$ with respect to the Lebesgue measure. Indeed, the Laplace transform of $U$ is given by $\mathcal{L}U(\lambda) = 1/\phi(\lambda)$, hence one can invoke the Tauberian and monotone density theorems to obtain the asymptotic behavior of $U$ from the asymptotic behavior of $\phi$.

We will be mainly interested in the behavior of the Green function $G(x, y)$ and the jumping function $J(x)$ near zero, hence the reasonable assumption on $\phi$ will be that it is regularly varying at infinity with index $\alpha \in [0, 1]$. This includes subordinators having a drift, as well as subordinators with slowly varying Laplace exponent at infinity, for example, a gamma subordinator. In the latter case we have $\phi(\lambda) = \log(1 + \lambda)$ and hence $\alpha = 0$.

## 1.2  Outline of the Book

Precise boundary estimates and explicit structure of $\alpha$-harmonic functions on general sub-domains of $\mathbb{R}^d$ remained for a long time essentially beyond the reach of the general theory. They are now objects of the **Boundary Potential Theory** which is discussed in Chapters 2 and 3 of the book. The main results of this theory: the **Boundary Harnack Principle**, the **3G Theorem**, the potential theory of **Schrödinger-type perturbations** and the **Conditional Gauge Theorem** are proven and discussed in Chapter 2. In Chapter 3 we present the important topic of nontangential limits of $\alpha$-harmonic functions on the border of the domain, whose main result is the **Relative Fatou Theorem** for $\alpha$-harmonic functions. Boundary Potential Theory for relativistic stable processes is also presented in Chapter 3.

The **Spectral Theory** of stable and related processes is an important tool of the Stochastic Potential Theory. It is the subject of Chapter 4 of the book. Its main results: Intrinsic Ultracontractivity, connection to Steklov problem, eigenvalue estimates, isoperimetric inequalities and estimates of the spectral gap are presented.

The second part of the book, contained in Chapter 5, is devoted to the **Potential Theory for Subordinate Brownian motions**, processes more general than $\alpha$–stable processes. Both classical Potential Theory and Boundary Potential Theory are presented for the subordinators and the subordinate processes. The main examples of subordinate processes covered by these results are stable and relativistic stable processes, geometric stable processes and iterated geometric stable processes. Important classes of gamma subordinators and Bessel subordinators are also included. In the last section of this chapter the underlying Brownian motion is replaced by the Brownian motion killed upon exiting a Lipschitz domain $D$.

The book has a form of extended lecture notes. We often strive to suggest ideas and relationships at the cost of the generality and completeness.

When the ideas are too cumbersome to verbalize, we choose to refer to the original papers. Occasionally, we present new results and modifications or simplifications of existing proofs. We often employ probabilistic notions and interpretations because they are extremely valuable for understanding of the ideas. They also lead to concise notation and powerful technical tools. Generally speaking, the main perspective that the probabilistic potential theory can offer is that of the distribution of the underlying process on the path space, as constructed by Kolmogorov. The distribution is a much richer object than the corresponding transition probability (or semigroup). The fundamental property of the distribution is the strong Markov property, first proved by Hunt for the Brownian motion. It should be noted that this property supersedes the Chapman-Kolmogorov equations allowing for the use of *random* stopping times–chiefly the first exit times of sub-domains of the state space with the natural hierarchy given by inclusion of the domains. For instance, the strong Markov property yields the mean value property for the Green potentials of measures off the support of these measures.

# Chapter 2
# Boundary Potential Theory for Schrödinger Operators Based on Fractional Laplacian

by K. Bogdan and T. Byczkowski

## 2.1 Introduction

Precise boundary estimates and explicit structure of harmonic functions are closely related to the so-called Boundary Harnack Principle (**BHP**). The proof of **BHP** for classical harmonic functions was given in 1977-78 by H. Dahlberg in [65], A. Ancona in [3] and J.-M. Wu in [153] (we also refer to [99] for a streamlined exposition and additional results). The results were obtained within the realm of the analytic potential theory. A probabilistic proof of **BHP**, one which employs only elementary properties of the Brownian motion, was given in [11]. The proof encouraged subsequent attempts to generalize **BHP** to other processes, in particular to the processes of jump type.

   **BHP** asserts that the ratio $u(x)/v(x)$ of nonnegative functions harmonic on a domain $D$ which vanish outside the domain near a part of the domain's boundary, $\partial D$, is bounded inside the domain near this part of $\partial D$. The result requires assumptions on the underlying Markov process and the domain. For Lipschitz domains and harmonic functions of the isotropic $\alpha$-stable Lévy process $(0 < \alpha < 2)$, **BHP** was proved in [27]. Another proof, motivated by [11], was obtained in [31] and extensions beyond Lipschitz domains were obtained in [150] and [38]. In particular the results of [38] provide a conclusion of a part of the research in this subject, and offer techniques that may be used for other jump-type processes.

   Lipschitz **BHP** leads to Martin representation of nonnegative $\alpha$-harmonic functions on Lipschitz domains ([28] and [56]). Another important consequence of **BHP** are sharp estimates of the Green function of Lipschitz domains and the so-called **3G** Theorem (see (2.26) below). We give these applications in the first part of the chapter, along with a self-contained proof of **BHP**, following [27] and [38].

   In the second part of the chapter we focus on the potential theory of Schrödinger-type perturbations, $\Delta^{\alpha/2} + q$, of the fractional Laplacian on subdomains of $\mathbb{R}^d$. The main result we discuss here is the Conditional Gauge Theorem (**CGT**), asserting comparability of the Green function of $\Delta^{\alpha/2} + q$

K. Bogdan et al., *Potential Analysis of Stable Processes and its Extensions*, Lecture Notes in Mathematics 1980, DOI 10.1007/978-3-642-02141-1_2, © Springer-Verlag Berlin Heidelberg 2009

with that of $\Delta^{\alpha/2}$, under an assumption of "non-explosion". Here $0 < \alpha < 2$, and the proof of **CGT** relies on the **3G** Theorem, thus on (Lipschitz) **BHP**. In presenting these results we generally follow the approach of papers [32] and [33]. The approach was modeled after [62], which deals with the Laplacian and its underlying process of the Brownian motion (see [64] for Schrödinger perturbations of elliptic partial differential operators of second order). For a different technique we refer to [54]. It should be noted that there are many algebraic similarities between the fractional Laplacian ($\alpha < 2$) and the Laplacian ($\alpha = 2$), but there are also deep analytical differences between these two cases, primarily due to the discontinuity of paths of the isotropic $\alpha$-stable Lévy process for $0 < \alpha < 2$.

## 2.2  Boundary Harnack Principle

Below we freely mix ideas from [27], [31], [32], [150], and [38], with some didactic improvements and modifications aimed at the simplification of presentation. In particular we give perhaps the shortest existing proof of **BHP** for $\alpha$-harmonic functions.

In what follows nonempty $D \subset \mathbb{R}^d$ is open. We intend to present the main ideas of the proof of **BHP** as given in [38] for arbitrary domains. However, for the simplicity of the discussion in the remainder of this chapter unless stated otherwise, *we will assume that $D$ is a Lipschitz domain*, and we will concentrate on *finite nonnegative functions $f$ on $\mathbb{R}^d$, which are represented on $D$ as Poisson integrals of their values on $D^c$*:

$$f(x) = \int_{D^c} f(y) P_D(x, y) dy, \quad x \in D^c. \tag{2.1}$$

For instance, if ($D$ is a Lipschitz domain and) $f \geqslant 0$ is bounded on $\overline{D}$, then $f = P_D[f]$ on $D$, see [27]. For a general discussion of the notion of $\alpha$-harmonicity we refer the reader to [32, 38]. We should perhaps state a warning that some aspects of the notion are richer and even counter-intuitive when confronted with the properties of harmonic functions of local operators. In particular, non-negativeness of functions which are $\alpha$-harmonic on $D$ is useful only if assumed on the whole of $\mathbb{R}^d$ (rather than merely on $D$). For instance, if $|y| > r$, then the function

$$B_r \ni x \mapsto \left[ \sup_{v \in B_r} P_{B_r}(v, y) \right] - P_{B_r}(x, y),$$

takes on the minimum of zero in an *interior* point of $B_r$, in stark contrast with the Harnack inequality. The reader may also want to consider (non-Lipschitz) domains with boundary of positive Lebesgue measure and domains

**Fig. 2.1** $D$, $B_r$, and outer cone

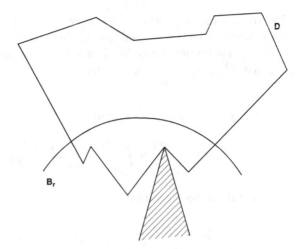

with complement of zero Lebesgue measure but positive Riesz capacity, to apprehend the complexity of the boundary problems for $\alpha$-harmonic functions.

For function $f \geqslant 0$ satisfying (2.1) we have $\Delta^{\alpha/2} f(x) = 0$ on $D$, see [32]. Furthermore, for *every* open $U \subset D$ we have

$$f(x) = \int_{U^c} f(y)\omega_U^x(dy), \quad x \in U. \tag{2.2}$$

This follows from (1.51). We emphasize that for the above *mean value property* of Poisson integrals it is *not* necessary that $\overline{U}$ be a *compact subset* of $D$, and we to refer the reader to [38] for cautions needed to deal with the general nonnegative $\alpha$-harmonic functions.

When $0 < r \leqslant 1$ we let $D_r = D \cap B_r$, a domain with the outer cone property, see Figure 2.1. We will often use (2.2) for $U = D_r$. We note that $\omega_{D_r}^x(\partial D_r) = 0$ for $x \in D_r$, in particular we can employ (1.53) for such $U$.

Consider $B = B_1$ and assume that

$$f = 0 \text{ on } B \setminus D. \tag{2.3}$$

Since $G_{D_r} \leqslant G_{B_r}$ (see (1.45), (1.46)), by the definition of Poisson kernel (1.49) we get

$$P_{D_r}(x, y) \leqslant P_{B_r}(x, y), \quad x \in D_r, y \in B_r^c.$$

By the mean value property and the assumption (2.3) we obtain

$$f(x) \leqslant \int_{B_r^c} f(y) P_{B_r}(x, y) dy, \quad x \in B_r, \quad 0 < r \leqslant 1. \tag{2.4}$$

The function $P_{B_r}(x, y)$ has a singularity at $|y| = r$. To remove this inconvenience, we will consider an analogue of volume averaging used on occasions in the classical potential theory. We fix a nonnegative function $\phi \in C_c^\infty((1/2, 1))$ such that $\int_{1/2}^1 \phi(r)\, dr = 1$ and we define

$$\psi(x, y) = \int_{1/2}^1 \phi(r)\, P_{B_r}(x, y)\, dr$$

$$= C_\alpha^d |y - x|^{-d} \int_{|y|\wedge 1/2}^{|y|\wedge 1} \frac{(r^2 - |x|^2)^{\alpha/2}}{(|y|^2 - r^2)^{\alpha/2}} \phi(r)\, dr, \quad x, y \in \mathbb{R}^d.$$

It is not difficult to check that

$$|\psi(x, y)| \leqslant \frac{C}{(1 + |y|)^{d+\alpha}}, \quad |x| \leqslant 1/3, \ y \in \mathbb{R}^d. \tag{2.5}$$

By Fubini's theorem and (2.4) we obtain

$$f(x) \leqslant \int_{B_r^c} f(y)\psi(x, y)dy \leqslant C \int_{\mathbb{R}^d} f(y)(1 + |y|)^{-d-\alpha}, \quad x \in B_{1/3}. \tag{2.6}$$

To obtain a reverse inequality for $x \in D_1 = D \cap B$ being not too close to $\partial D_1$ we note that $P_{B_r}(0, y) \geqslant C_\alpha^d r^\alpha |y|^{-d-\alpha}$, see (1.57). If $r_0 > 0$ and $B(2r_0, x_0) \in D_1$, then

$$f(x_0) = \int_{B^c(x_0, r_0)} P_{r_0}(0, y - x_0)f(y)dy \geqslant \int_{B^c(x_0, r_0)} C_\alpha^d r_0^\alpha |y - x_0|^{-d-\alpha} f(y)dy. \tag{2.7}$$

By the Harnack inequality for $f$ on $B(x_0, r_0)$ we can enlarge the domain of integration so that

$$f(x_0) \geqslant c \int_{\mathbb{R}^d} (1 + |y|)^{-d-\alpha} f(y)dy.$$

Here and in what follows the *constants* ($c$, $C$ etc.) depend on $d$, $\alpha$ and $D$, in particular on $r_0$.

This and (2.6) yield the following Carleson-type estimate.

**Corollary 2.1.** *There is a constant $C$ depending only on $d$, $\alpha$, and $x_0$ such that*

$$f(x) \leqslant Cf(x_0), \quad x, x_0 \in D_{1/3}. \tag{2.8}$$

In what follows we will consider $D_{1/4}$ and will fix $x_0 \in D_{1/5}$. We have

$$f(x) = \int_{D_{1/4}^c} f(y)P_{D_{1/4}}(x, y)dy = \int_{D_{1/4}} G_{D_{1/4}}(x, v)\kappa(v)dv, \tag{2.9}$$

where
$$\kappa(v) = \int_{D^c_{1/4}} \mathcal{A}_{d,-\alpha}|y - v|^{-d-\alpha}f(y)dy\,, \quad v \in D_{1/4}\,.$$

We thus have $f$ expressed as the Green potential of the charge $\kappa(v)$ interpreted as the intensity of jumps of $Y$ "to" $f$ on $D^c$. Let

$$\kappa_1(v) = \int_{B^c_{1/3}} \mathcal{A}_{d,-\alpha}|y - v|^{-d-\alpha}f(y)dy\,, \quad v \in D_{1/4}\,,$$

$$\kappa_2(v) = \int_{B_{1/3}\setminus D_{1/4}} \mathcal{A}_{d,-\alpha}|y - v|^{-d-\alpha}f(y)dy\,, \quad v \in D_{1/4}\,,$$

and
$$f_i(x) = \int_{D_{1/4}} G_{D_{1/4}}(x,v)\kappa_i(v)dv\,, \quad i = 1, 2\,. \tag{2.10}$$

We note that $f_i$ are $\alpha$-harmonic, in fact Poisson integrals, on $D_{1/4}$. We observe that $\kappa_1$ is bounded, in fact *nearly constant* on $D_{1/4}$:

$$c^{-1}\kappa_1(v_2) \leqslant \kappa_1(v_1) \leqslant c\kappa_1(v_2)\,, \quad v_1, v_2 \in D_{1/4}\,, \tag{2.11}$$

because $|y - v|^{-d-\alpha}$ is nearly constant in $v \in D_{1/4}$ (uniformly in $y \in B^c_{1/3}$). Also, $\kappa_1(v) \leqslant cf(x_0)$, see (2.7). Thus

$$f_1(x) \leqslant cf(x_0)\int_{D_{1/4}} G_{D_{1/4}}(x,v)dv = cf(x_0)s_{D_{1/4}}(x)\,, \quad x \in D_{1/4}\,. \tag{2.12}$$

We will see that $s_{D_{1/4}}$ faithfully represents the asymptotics of $f = f_1 + f_2$ at $\partial D \cap B_{1/5}$. To this end we first note that by (2.8),

$$f_2(x) \leqslant Cf(x_0)\omega^x_{D_{1/4}}(B^c_{1/4})\,, \quad x \in D_{1/4}\,. \tag{2.13}$$

**Lemma 2.2.** *For every $p \in (0, 1)$ there is a constant $C$ such that if $D \subset B$ then*
$$\omega^x_D(B^c) \leqslant C\, s_D(x)\,, \quad x \in D_p\,.$$

*Proof.* Let $0 < p < 1$. We choose a function $\varphi \in C^\infty_c(\mathbb{R}^d)$ such that $0 \leqslant \varphi \leqslant 1$, $\varphi(y) = 1$ if $|y| \leqslant p$, and $\varphi(y) = 0$ if $|y| \geqslant 1$. Let $x \in D_p$. By (1.47) we have

$$\omega^x_D(B^c) = \int_{B^c} (\varphi(x) - \varphi(y))\omega^x_D(dy) \leqslant \int_{D^c} (\varphi(x) - \varphi(y))\omega^x_D(dy)$$

$$= -\int_D G_D(x,y)\Delta^{\alpha/2}\varphi(y)dy\,.$$

It remains to observe that $\Delta^{\alpha/2}\varphi$ is bounded and the lemma follows.   $\square$

By (2.13), scaling and Lemma 2.2 (with $p = 4/5$) we obtain that $f_2(x) \leqslant cf(x_0)s_{D_{1/4}}(x)$ for $x \in D_{1/5}$. This, and (2.12) yield the following improvement of Carleson estimate

$$c^{-1}f(x_0)s_{D_{1/4}}(x) \leqslant f(x) \leqslant cf(x_0)s_{D_{1/4}}(x), \quad x \in D_{1/5}. \tag{2.14}$$

Indeed, the lower bound in (2.14) follows from the inequality

$$f(x) \geqslant \int_{D_{1/4}} G_{D_{1/4}}(x,v)\kappa_3(v)dv,$$

where

$$\kappa_3(v) = \int_{B(x',r')} f(y)\mathcal{A}_{d,-\alpha}|y-v|^{-d-\alpha}dy \geqslant cf(x_0), \quad v \in D_{1/4},$$

and $B(2r',x') \subset D_{1/4} \setminus D_{1/5}$ is a ball (if the set $D_{1/4} \setminus D_{1/5}$ is empty then $f_2 \equiv 0$, and we simply use (2.11) and (2.10)).

The following Boundary Harnack Principle is a direct analogue of (1.14).

**Theorem 2.3 (BHP).** *If functions $f_1$ and $f_2$ satisfy the above assumptions on $f$, then*

$$f_1(x)f_2(y) \leqslant Cf_1(y)f_2(x), \quad x,y \in D_{1/5}.$$

*Proof.* We fix $x_0 \in D \cap B_{1/5}$. For $x,y \in D \cap B_{1/5}$ we obtain from (2.14)

$$f_1(x)f_2(y) \leqslant c^2 f_1(x_0)f_2(x_0)s_{D_{1/4}}(x)s_{D_{1/4}}(y),$$

and

$$f_1(y)f_2(x) \geqslant c^{-2}f_1(x_0)f_2(x_0)s_{D_{1/4}}(y)s_{D_{1/4}}(x).$$

The result, translation and scaling invariance of the class of $\alpha$-harmonic functions, and the usual Harnack inequality, allow to estimate the growth of $\alpha$-harmonic functions vanishing at a part of the domain's boundary up to this part of the boundary. The constant $C$ in our present proof depends on $D$ (and the choice of $x_0$), however a more delicate and technical proof shows that $C$ may be so chosen to depend *only* on $d$ and $\alpha$. We refer the reader to [38] for this important strengthening of **BHP**. An important consequence of the domain-independent, or *uniform* **BHP** of [38] is given in the following statement

$$\lim_{D \ni x \to 0} f_1(x)/f_2(x) \quad \text{exists.} \tag{2.15}$$

**BHP** and (2.15) were given in [27] (see also [31]) for Lipschitz domains, generalized in [150] to the so-called $\kappa$-fat domains, and proved for *arbitrary* open sets in [38]. The proof of (2.15) seems too technical to be discussed here, but we will hopefully give some insight into its main idea, when discussing the uniqueness of the Martin kernel with the pole at infinity for cones.

Let $\rho(x) = \text{dist}(x, D^c)$. Compared to **BHP**, the following local estimate for individual (nonnegative) Poisson integral on Lipschitz domains, if not sharp, is more explicit.

**Lemma 2.4.** *Let $\Gamma : \mathbb{R}^d \to \mathbb{R}$ satisfy (1.12), and $\Gamma(0) = 0$. Let $D = D_\Gamma \cap B$, and $A = (0, 0, \ldots, 0, 1/2) \in D$. There are $C = C(d, \alpha, \lambda)$ and $\epsilon = \epsilon(d, \alpha, \lambda) \in (0, \alpha)$ such that*

$$C^{-1} f(A) \rho(x)^{\alpha - \epsilon} \leqslant f(x) \leqslant C f(A) \rho(x)^\epsilon, \quad x \in D_{1/2}. \tag{2.16}$$

The right hand side of (2.16) is a strengthening of the Carleson estimate, and it asserts a power-type decay of $u$ at the boundary of $D$. This decay rate is related to the existence of outer cones for the boundary points of $D$, and steady escape of mass of the process when it approaches $\partial D$ (see our discussion above of the fact that $\omega_D^x(\partial D) = 0$). For a class of domains including domains with the boundary defined by a $C^2$ function we have $\epsilon = \alpha/2$, which may be verified by a direct calculation involving the Green function of the ball, and of the complement of the ball, see [56], [109]. Then (2.16) becomes *sharp*, meaning that all sides of the inequality are in fact *comparable*. The exponent $\alpha/2$ is also related to the fact that

$$f(x) = x_+^{\alpha/2}, \quad x \in \mathbb{R}, \tag{2.17}$$

is $\alpha$-harmonic on the half-line $\{x > 0\}$, see [30] for explicit calculations involving $\Delta^{\alpha/2}$.

For general Lipschitz domains the exponent $\epsilon$ on the right-hand side of (2.16) is usually not given explicitly. We like to note that $\epsilon > 0$ may be arbitrarily small, e.g. for the complement of cone with sufficiently small opening in dimension $d > 2$. For a more detailed study of the asymptotic behavior of $\alpha$-harmonic functions in cones, and some open problems we refer the reader to [5] and [123].

We will briefly discuss the left hand side inequality in (2.16). We like to emphasize the fact that the power-type decay cannot be arbitrarily fast, a significant difference when compared with the classical harmonic functions in narrow cones. Indeed, $\epsilon > 0$ may be arbitrarily small (for very narrow cones), but we always have $\alpha - \varepsilon < \alpha$! This is a noteworthy contrast with the classical potential theory ($\alpha = 2$). For an explanation of this phenomenon we will consider exponentially shrinking disjoint balls $B^k = B(A^k, cr^k)$, where $0 < r < 1$ and $c$ are such that $B^k \subset D_{r^k}$ ($k = 0, 1, \ldots$), see Figure 2.2. By the mean value property, we have

$$f(A^k) \geqslant \sum_{l=0}^{k-1} \int_{B^l} f(y) \omega_{B^k}^{A^k}(dy) \tag{2.18}$$

$$\geqslant C \sum_{l=0}^{k-1} \int_{B^l} f(A^l) r^{(k-l)\alpha},$$

**Fig. 2.2** Exponentially
shrinking balls

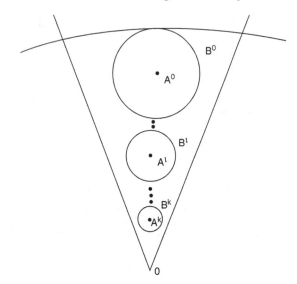

where we used the formula for the Poisson kernel of the ball. Thus, $\beta_k :=$ $f(A^k)r^{-k\alpha} \geqslant C\sum_{l=0}^{k-1}\beta_l$. By induction we see that $\beta_k \geqslant C(1+C)^k\beta_0$, which yields the exponent $\alpha - \varepsilon < \alpha$ on the left hand side of (2.16).

We note that the first term of the sum in (2.18) approximately equals $r^{k\alpha}f(A_0)$, which is much smaller than the whole sum. Thus a direct jump (say, to $B_0$) has a negligible impact on the values of the $\alpha$-harmonic function on $B^k$. Instead, the many combined shorter jumps between the balls $\{B^l\}$ yield the main contribution. The geometry of Lipschitz domains plays a role here. Domains which are "thinner" at some boundary points may show a different decay rate of $\alpha$-harmonic functions (i.e. that given by a few direct jumps may prevail, see [125]). This observation leads to a notion of *inaccessibility* developed in [38].

We want to point out after [38], that **BHP** can be studied as a property of the Poisson kernel and the Green function, without even referring to the notion of $\alpha$-harmonicity. In fact, the main application of **BHP** is the following one, to $f_1(x) = G_D(x, x_1)$ and $f_2(x) = G_D(x, x_2)$, for $x$ (in a Lipschitz subset of) $D \setminus \{x_1, x_2\}$. We fix an arbitrary reference point $x_0 \in D$ and we define the Martin kernel of $D$,

$$M_D(x, y) = \lim_{D \ni v \to y} \frac{G_D(x, v)}{G_D(x_0, v)}, \quad x \in \mathbb{R}^d, \ y \in \partial D. \qquad (2.19)$$

**Theorem 2.5.** *The limit in (2.19) exists. $x \mapsto M_D(x, y)$ is up to constant multiples the only nonnegative $\alpha$-harmonic function on $D$ and equal to zero on $D^c$ which continuously vanishes at $D^c \setminus \{y\}$.*

The existence part of the result follows easily from (2.15). The $\alpha$-harmonicity of $M_D$, however, depends delicately on the Lipschitz geometry of the domain via the lower bound in (2.16), see [28]. We refer the reader to [28] for an elementary study of the properties of $M_D(\cdot, y)$ for Lipschitz domains. We also refer to [38] for the case of arbitrary open set and for the explanation of the role played by the *accessibility* of the point $y$ from within the set.

It should be noted that $M_D(\cdot, y)$ is *not* of the form (2.1). Nonnegative $\alpha$-harmonic functions *vanishing* on $D^c$ are called *singular $\alpha$-harmonic*. They resemble classical Poisson integrals of singular measures on the sphere (and also nonnegative martingales converging to zero almost surely).

We will cite after [28] the representation theorem for nonnegative $\alpha$-harmonic functions on bounded Lipschitz domains $D$ (for arbitrary nonempty open subsets of $\mathbb{R}^d$ see [38]).

**Theorem 2.6.** *For every function $u \geqslant 0$ which is $\alpha$-harmonic in $D$ there exists a unique finite measure $\mu \geqslant 0$ on $\partial D$, such that*

$$u(x) = \int_{D^c} P_D(x, y) u(y) dy + \int_{\partial D} M_D(x, y) \mu(dy), \quad x \in D. \tag{2.20}$$

In view of the recent developments in [38] we like to make the following comments. First, $\int_{D^c} P_D(x, y) u(y) dy$ above may be generalized to Poisson integrals of nonnegative *measures*:

$$\int_{D^c} P_D(x, y) \lambda(dy) < \infty, \tag{2.21}$$

and it is legitimate to regard $D^c$ as the "Martin boundary" of (bounded Lipschitz) $D$ for $\Delta^{\alpha/2}$, with kernels $M_D(\cdot, y)$, $y \in \partial D$, and $P_D(\cdot, y) + \delta_y(\cdot)$, $y \in D^c \setminus \partial D$. Second, for *general* domains in arbitrary dimension, *inaccessible* points of the Euclidean boundary will contribute a Poisson kernel, rather than a Martin kernel. Third, for unbounded domains a Martin kernel may be attributed to the point at infinity (if accessible). For details we refer the reader to [38], which appears to finalize the problem of representing nonnegative $\alpha$-harmonic functions, and offers notions and methods appropriate for handling more general Markov processes with jumps. To further encourage the interested reader, we want to point out that for bounded domains their "Martin boundary" decreases when the domain increases [38]. Comparing to $\Delta$, we see that the potential theory of $\Delta^{\alpha/2}$ is more compatible with the Euclidean topology of $\mathbb{R}^d$.

We return to considering a Lipschitz domain $D \subset \mathbb{R}^d$ in dimension $d \geqslant 2$. For $y \in \partial D$, $M_D(x, y)$ is (up to constant multiples) the unique $\alpha$-harmonic function continuously vanishing on $D^c \setminus \{y\}$ (and having a singularity at $y$, which "feeds" the function through (1.63)). As remarked above, a similar function can be constructed for the point at infinity, if $D$ is unbounded:

$$M(x) = M_D(x, \infty) = \lim_{D \ni v, |v| \to \infty} \frac{G_D(x, v)}{G_D(x_0, v)}, \quad x \in \mathbb{R}^d. \qquad (2.22)$$

In the case when $D$ is an open *cone* $\mathcal{C} \subset \mathbb{R}^d$, the existence, uniqueness and homogeneity properties of $M$ were studied [5] and [123]. Below we will give a flavor of the technique used in the study. We first note that the mean value property holds for such $M$ for *every bounded* open subset $U$ of $\mathcal{C}$, as the pole is so far away. Let $\mathbf{1} \neq 0$ be a point in $\mathbb{R}^d$ (say $\mathbf{1} = (0, \ldots, 0, 1)$). For $x \in \mathbb{R}^d \backslash \{0\}$, we denote by $\theta(x)$ the angle between $x$ and $\mathbf{1}$. The right circular cone of angle $\Theta \in (0, \pi)$ is the Lipschitz domain

$$\mathcal{C} = \mathcal{C}_\Theta = \{x \in \mathbb{R}^d : \theta(x) < \Theta\}.$$

Clearly, for every $r > 0$ we have $r\mathcal{C} = \mathcal{C}$. In particular, by scaling, if $u$ is $\alpha$-harmonic on $\mathcal{C}$, then so is $x \mapsto u(rx)$. We will prove the uniqueness of $M$. To this end, we assume that there is another function $m \geqslant 0$ on $\mathbb{R}^d$ which vanishes on $\mathcal{C}^c$, satisfies $m(\mathbf{1}) = 1$ and

$$m(x) = \mathbb{E}_x m(Y_{\tau_B}), \quad x \in \mathbb{R}^d,$$

for every open bounded $B \subset \mathcal{C}$. By **BHP**,

$$C^{-1} m(x) \leqslant M(x) \leqslant C m(x),$$

for $x \in B \cap \mathcal{C}$. By scaling, this extends to all $x \in \mathcal{C}$ with the same constant. We let $a = \inf_{x \in \mathcal{C}} m(x)/M(x)$. For clarity, we note that $C^{-1} \leqslant a \leqslant 1$. Let $R(x) = m(x) - aM(x)$, so that $R \geqslant 0$ on $\mathbb{R}^d$. Assume (falsely) that $R(x) > 0$ for some, and therefore for every $x \in \mathcal{C}$. Then, by **BHP** and scaling,

$$R(x) \geqslant \varepsilon M(x), \quad x \in \mathbb{R}^d,$$

for some $\varepsilon > 0$. We have

$$a = \inf_{x \in \mathcal{C}} \frac{m(x)}{M(x)} = \inf_{x \in \mathcal{C}} \frac{aM(x) + R(x)}{M(x)} \geqslant a + \varepsilon,$$

which is a contradiction. Thus $R \equiv 0$, $m = aM$, and the normalizing condition $m(\mathbf{1}) = M(\mathbf{1}) = 1$ yields $a = 1$. The uniqueness of $M$ is verified.

We like to note that the existence of the limits of the ratios of nonnegative $\alpha$-harmonic functions, (2.15), is proved by a similar argument, see [27, 38]. This *oscillation-reducing* mechanism of **BHP** is well known for local operators, e.g. Laplacian ([11]), but the non-local character of the fractional Laplacian seriously complicates such arguments, except in some special cases, like that of the cone. Some elements of the proof (of vanishing of oscillations of ratios of non-negative $\alpha$-harmonic functions) are given in [27]. The complete details in the generality of arbitrary domains are given in [38].

To appreciate the importance of uniqueness, we return to the discussion of the Martin kernel with the pole at infinity for the cone. By scaling, for every $k > 0$ the function $M(kx)/M(k\mathbf{1})$ satisfies the hypotheses defining $M$. Thus it is equal to $M$, or

$$M(kx) = M(x)M(k\mathbf{1}) \quad x \in \mathbb{R}^d \,.$$

In particular, $M(kl\mathbf{1}) = M(l\mathbf{1})M(k\mathbf{1})$ for positive $k, l$. By continuity of $\alpha$-harmonic functions on the domain of harmonicity, there exists $\beta$ such that $M(k\mathbf{1}) = k^{\beta}M(\mathbf{1}) = k^{\beta}$ and

$$M(kx) = k^{\beta}M(x), \quad x \in \mathbb{R}^d \,,$$

or

$$M(x) = |x|^{\beta}M(x/|x|), \quad x \neq 0 \,, \tag{2.23}$$

compare (2.16). By (2.14), $M$ is locally bounded and tends to zero at the origin, thus

$$0 < \beta < \alpha \,. \tag{2.24}$$

It is known that $\beta$ is close to $\alpha$ for very narrow cones, and it will be close to 0 for obtuse cones (for $\Theta$ close to $\pi$), at least in dimension $d \geqslant 2$. We refer the reader to [5], [123], [35] for more information and a few explicit values of $\beta$ for specific cones (see (2.17) for the half-line).

## 2.3  Approximate Factorization of Green Function

In this section we will consider a *bounded* Lipschitz domain $D \subset \mathbb{R}^d$, $d \geqslant 2$, with Lipschitz constant $\lambda$. To simplify formulas, we recall the notation $\approx$: we write $f(y) \approx g(y)$ for $y \in A$ if there exist constants $C_1, C_2$ not depending on $y$ such that $C_1 f(y) \leqslant g(y) \leqslant C_2 f(y)$, $y \in A$.

Let $\delta(x) = \operatorname{dist}(x, D^c)$. We fix $x_0, x_1 \in D$, $x_0 \neq x_1$, and we let $\kappa = 1/(2\sqrt{1 + \lambda^2})$. For $x, y \in D$ we denote $r = r(x, y) = \delta(x) \vee \delta(y) \vee |x - y|$. For *small* $r > 0$ we write $\mathcal{B}(x, y)$ for the set of points $A$ such that $B(A, \kappa r) \subset D \cap B(x, 3r) \cap B(y, 3r)$, see Figure 2.3, and we put $\mathcal{B}(x, y) = \{x_1\}$ for *large* $r$ [29]. The set $\mathcal{B}(x, y)$ is nonempty (see [98] or [29] for details). Informally speaking, $A \in \mathcal{B}(x, y)$ dominates $x$ and $y$ similarly as $A$ of Lemma 2.4 dominates the points of $D_{1/2}$, see Figure 2.3. Let $G = G_D$, the Green function of $D$ for the fractional Laplacian. We define

$$\phi(x) = G(x_0, x) \wedge c \,.$$

**Fig. 2.3** $A \in \mathcal{B}(x, y)$

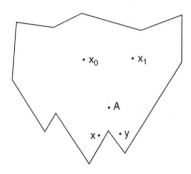

The following is a sharp, if not completely explicit, *approximate factorization* of $G(x, y)$.

$$C^{-1} \frac{\phi(x)\phi(y)}{\phi^2(A)} |x - y|^{\alpha-d} \leqslant G(x, y) \leqslant C \frac{\phi(x)\phi(y)}{\phi^2(A)} |x - y|^{\alpha-d} . \qquad (2.25)$$

Here $A$ is an arbitrary point of $\mathcal{B}(x, y)$. A proof of (2.25) is given in [98] (see also [29] for the case of $\alpha = 2$). We will sketch the proof.

If $x$ and $y$ are close to each other but far from the boundary, then (2.25) is equivalent to $G(x, y) \approx |x - y|^{\alpha-d}$, because the term subtracted in (1.45) is small.

Another case to consider is the situation of $|y - x|$ being large and $\delta(x)$, $\delta(y)$ being small. By symmetry, $G(x, y)$ is $\alpha$-harmonic both in $x$, and in $y$ (on $D \setminus \{y\}$ and $D \setminus \{x\}$, correspondingly). By **BHP** (and the usual Harnack inequality) $G(x, y)/G(x_0, y) \approx G(x, x_1)/G(x_0, x_1)$. Since $0 < G(x_0, x_1) < \infty$ is a constant, we obtain (2.25) in the considered case. If $|y - x|$, $\delta(x)$, and $\delta(y)$ are all small then we use **BHP** in a similar way, but *twice*. If $\delta(x)$ is small, and $\delta(y)$ is large, then $G(x, y) = G(y, x) \approx G(x_0, x)$ by the Harnack inequality.

We remark that $\phi(\cdot)$ may be replaced by $s_D(\cdot)$ in (2.25), compare Lemma 2.2, [38]. For bounded $C^2$ domains we may use $\phi(\cdot) = \delta^{\alpha/2}(\cdot)$, obtaining an estimate which is both sharp and explicit, [109, 56].

We will give a short proof of the following celebrated inequality known as **3G** Theorem:

**Theorem 2.7 (3G).**

$$\frac{G(x, y)G(y, z)}{G(x, z)} \leqslant C \frac{|x - y|^{\alpha-d}|y - z|^{\alpha-d}}{|x - z|^{\alpha-d}}, \quad x, y, z \in D. \qquad (2.26)$$

*Proof.* let $x, y, z \in D$ and $R \in \mathcal{B}(x, y)$, $S \in \mathcal{B}(y, z)$, $T \in \mathcal{B}(x, z)$. By (2.25),

$$\frac{G(x, y)G(y, z)}{G(x, z)} \leqslant C \frac{|x - y|^{\alpha-d}|y - z|^{\alpha-d}}{|x - z|^{\alpha-d}} W^2,$$

where $W = [\phi(y)\phi(T)]/[\phi(R)\phi(S)]$. We will verify the boundedness of $W$. Let $r_1 = \delta(x) \vee \delta(y) \vee |x-y|$, $r_2 = \delta(y) \vee \delta(z) \vee |y-z|$, and $r_3 = \delta(x) \vee \delta(z) \vee |x-z|$ because $\phi$ is bounded. If $R = x_1$ and $S = x_1$ then $W \leqslant C$. If $R \neq x_1$, that is if $r_1$ is small, then we choose $Q \in \partial D$ such that $\delta(y) = |y-Q|$. By the Carleson estimate $\phi(y) \leqslant C\phi(R)$. Consequently, if $S = x_1$, then $W \leqslant C$. The same holds true if $S \neq x_1$ and $R = x_1$. By symmetry, to complete the proof, we may assume that $r_1 \leqslant r_2$ are *small*. We have $r_3 \leqslant r_1 + r_2 \leqslant 2r_2$, so $r_3$ is also small. In fact $|T-Q| \leqslant |T-z|+|y-z|+|y-Q| < 3r_3+r_2+r_2 \leqslant 8r_2$, therefore by the Carleson estimate and the Harnack inequality $\phi(T) \leqslant C\phi(S)$. Recall that $\phi(y) \leqslant C\phi(R)$, thus $W$ is bounded in this case, too. This finishes the proof.                                                                           □

Since $|x-y|^{\alpha-d}|y-z|^{\alpha-d}/|x-z|^{\alpha-d} \leqslant 2^{d-\alpha}\left(|x-y|^{\alpha-d}+|y-z|^{\alpha-d}\right)$, we obtain the following version of **3G**:

$$\frac{G(x,y)G(y,z)}{G(x,z)} \leqslant C\left(K_\alpha(x-y) + K_\alpha(y-z)\right), \quad x,y \in D. \qquad (2.27)$$

The definition of the Martin kernel yields

$$\frac{G(x,y)M_D(y,\xi)}{M_D(x,\xi)} \leqslant C\left(K_\alpha(x-y) + K_\alpha(y-\xi)\right), \quad x,y \in D, \, \xi \in \partial D.$$
$$(2.28)$$

As we have mentioned, the importance of **3G** in potential theory was observed in [64]. Below we will use a probabilistic framework of conditional processes to employ **3G** to construct and estimate the Green function of Schrödinger perturbations of $\Delta^{\alpha/2}$. Before we take our chances in this endeavor, however, we like to notice that a purely analytic approach to this problem also exists. The approach is based on the so called *perturbation series*, or Duhamel's formula, whose application is greatly simplified *if* **3G** is satisfied. We refer the reader to a self-contained exposition of this technique in [85] (see also [84]). Analogous consideration based on so-called *3P Theorem* of [36] yields comparability of the perturbed transition density with the original one. We refer the interested reader to [36] and [37] for these developments.

## 2.4   Schrödinger Operator and Conditional Gauge Theorem

We will focus on the potential theory of Schrödinger operators, $u \mapsto \Delta^{\alpha/2}u + qu$, on subdomains of $\mathbb{R}^d$, following the development of [32, 33, 62]. The class of admissible "potentials" $q$ is tailor-made for the transition probability of $\{Y_t\}$ (and $\Delta^{\alpha/2}$).

Put in a general perspective we consider here "small" additive perturbations of the generator of a semigroup and we expect the potential-theoretic object to be similar before and after these perturbations. In particular, the conditional gauge function defined below is the ratio between the Green functions after and before the perturbation, and the Conditional Gauge Theorem (**CGT**) asserts that the function is bounded under certain assumptions. Originally, many authors considered the Laplace operator and bounded $q$, or $q$ in a Kato class and smooth domains $D$, see the references in [64]. The paper [64] made an essential progress by including Lipschitz domains in the case of the Laplace operator. This direction of research is summarized in [62].

The paper [54] initiated in 1997 the study of **CGT** for rotation invariant stable Lévy and more general processes for Schrödinger and more general perturbations. The focus of [54] was on $C^{1,1}$ domains. **CGT** for the stable proceses in Lipschitz domains was proved in 1999 in [32]. We also like to note that there is a recent non-probabilistic approach to **CGT**, see [85] and [37].

A (Borel) function $q$ on $\mathbb{R}^d$ is said to belong to the *Kato class* $\mathcal{J}_\alpha$ if

$$\lim_{t\downarrow 0} \sup_{x\in\mathbb{R}^d} \int_0^t E^x |q(Y_s)| \, ds = 0. \tag{2.29}$$

Thus, (2.29) is a statement of *negligibility* of $q$ in (small) time, with respect to the given transition probability. To make (2.32) more explicit we recall the following well-known estimate (see (1.29)):

$$C^{-1}\left(\frac{t}{|x|^{d+\alpha}} \wedge t^{-d/\alpha}\right) \le p_t(x) \le C\left(\frac{t}{|x|^{d+\alpha}} \wedge t^{-d/\alpha}\right). \tag{2.30}$$

The estimate is proved by subordination (see, e.g., [39]). Noteworthy,

$$\frac{t}{|x|^{d+\alpha}} \le t^{-d/\alpha} \quad \text{iff} \quad t \le |x|^\alpha. \tag{2.31}$$

We easily see that

$$\int_0^t p_s(x)ds \approx \frac{t^2}{|x|^{d+\alpha}} \wedge \frac{1}{|x|^{\alpha-d}}.$$

By the definition of $E^x$, and Fubini-Tonelli, $q \in \mathcal{J}_\alpha$ if and only if

$$\lim_{t\downarrow 0} \sup_{x\in\mathbb{R}^d} \int_{\mathbb{R}^d} \left[\frac{t^2}{|y-x|^{d+\alpha}} \wedge \frac{1}{|y-x|^{\alpha-d}}\right] |q(y)|dy = 0.$$

It follows that (2.29) is equivalent to the following condition of *negligibility* of $q$ in (small) space with respect to the potential operator:

$$\lim_{\gamma \downarrow 0} \sup_{x \in \mathbb{R}^d} \int_{|x-y| \leq \gamma} K_\alpha(x - y) \, |q(y)| \, dy = 0 \,. \tag{2.32}$$

We also note that $L^\infty(\mathbb{R}^d) \subset \mathcal{J}_\alpha \subset L^1_{loc}(\mathbb{R}^d)$.

For $q \in \mathcal{J}_\alpha$ we define the *additive functional*

$$A_q(t) = \int_0^t q(Y_s) \, ds \,,$$

and the corresponding *multiplicative functional*

$$e_q(t) = \exp(A_q(t)) \,.$$

We have

$$e_q(t + s) = e_q(t) \, (e_q(s) \circ \theta_t) \,, \quad t, s \geq 0 \,.$$

Here $\theta_t$ is the usual shift operator acting on the process $Y$ by the formula: $Y_s \circ \theta_t = Y_{t+s}$.

For an open bounded set $D$ we define the killed Feynman-Kac semigroup $T_t$ by the formula

$$T_t f(x) = E^x[t < \tau_D; e_q(t) \, f(Y_t)] \,. \tag{2.33}$$

$T_t$ is a strongly continuous semigroup on $L_p(D), 1 \leq p < \infty$, and on $C(D)-$ for regular $D$. For each $t > 0$, the operator $T_t$ is determined by a symmetric transition density function $u_t(x, y)$ which is in $C_0(D \times D)$ for regular $D$. We should note that $\{T_t, \ t > 0\}$ is generated by $\Delta^{\alpha/2} + q$, see [32]. The next lemma is fundamental in the theory of Feynman-Kac semigroups– this is seen in the development of [62], which we will follow quite closely below.

**Lemma 2.8.** [Khasminski lemma] *Let $\tau$ be an optional time of $Y$ such that*

$$\tau \leq t + \tau \circ \theta_t, \quad \text{on } \{t < \tau\}, \, t > 0 \,. \tag{2.34}$$

*Suppose that $q \geq 0$ and $E^x A(\tau) < \infty$ for all $x \in \mathbb{R}^d$. Then for each integer $n \geq 0$ we have,*

$$\sup_x E^x[A(\tau)^n] \leq n! \sup_x (E^x A(\tau))^n \,. \tag{2.35}$$

*If $\sup_x E^x A(\tau) = \alpha < 1$ then*

$$\sup_x E^x e^{A(\tau)} \leq (1 - \alpha)^{-1} \,.$$

The condition (2.34) is satisfied if $\tau$ is constant or if $\tau = \tau_D$ for some $D \subseteq \mathbb{R}^d$.

*Proof.* Since $q \geqslant 0$, the functional $A(\cdot)$ is nonnegative and nondecreasing. By Fubini-Tonelli and (2.29), $A(\tau) < \infty$ a.s. We have

$$\frac{A(\tau)^{n+1}}{n+1} = \int_0^\tau [A(\tau) - A(t)]^n \, dA(t) \, .$$

For $t < \tau$, by (2.34),

$$A(\tau) - A(t) \leqslant A(t + \tau \circ \theta_t) - A(t) = \int_t^{t+\tau \circ \theta_t} q(Y_s) \, ds$$

$$= [\int_0^\tau q(Y_s) \, ds] \circ \theta_t = A(\tau) \circ \theta_t \, .$$

By Fubini's theorem,

$$\frac{E^x[A(\tau)^{n+1}]}{n+1} \leqslant E^x[\int_0^\tau [A(\tau) \circ \theta_t]^n \, dA(t)] = \int_0^\infty E^x[t < \tau; [A(\tau) \circ \theta_t]^n q(Y_t)] \, dt \, .$$

By the Markov property the last integral is equal to

$$\int_0^\infty E^x[t < \tau; E^{Y_t}[A(\tau)^n] \, q(Y_t)] \, dt \leqslant \sup_x E^x[A(\tau)^n] \int_0^\infty E^x[t < \tau; q(Y_t)] \, dt$$

$$= \sup_x E^x[A(\tau)^n] \, E^x A(\tau) \, .$$

It follows that

$$\sup_x E^x[A(\tau)^{n+1}] \leqslant (n+1) \sup_x E^x[A(\tau)^n] \sup_x E^x[A(\tau)] \, ,$$

hence (2.35) is proved by induction on $n$. The last assertion of the lemma is an immediate consequence of (2.35).                                                $\square$

We like to note that $A(t)$ increases where $q > 0$, and $e_q(t)$ may be interpreted as the *mass* of a particle moving along the trajectories of the process in the *potential well* given by $q$. If $q \leqslant 0$ then the mass is always bounded by 1 (subprobabilistic), which corresponds to Courrège's theorem, see (1.25).

The *gauge function* of $D$ and $q$ is defined as follows:

$$u(x) = E^x e_q(\tau_D) \, .$$

We can interpret $u(x)$ as the expected mass of the particle when it leaves the domain. We note that since $\tau_D$ is an unbounded random variable, the mass may be infinite if $q$ is (say, positive and) large enough. When the gauge function satisfies $u(x) < \infty$ for (some, hence for all) $x \in D$, we call the pair $(D, q)$ *gaugeable*.

We consider $u_t(x, y) = E^x[1_{t < \tau_D} e_q(t) | Y_t = y]$, the integral kernel of $T_t$. We define the Green function of the Schrödinger operator on $D$,

$$V(x, y) = \int_0^\infty u_t(x, y) \, dt \, .$$

The potential operator of the the Feynman-Kac semigroup $T_t$ killed off $D$ is, by definition

$$Vf(x) = \int_0^\infty T_t f(x) \, dt = \int_0^\infty E^x[t < \tau_D; e_q(t) f(Y_t)] \, dt$$
$$= E^x \int_0^{\tau_D} e_q(t) f(Y_t) \, dt = \int_D V(x, y) f(y) \, dy \, .$$

Both functions $V$ and $u_t$ are symmetric in $(x, y) \in D \times D$ and $u_t$ is continuous whenever $D$ is regular.

The theorem below provides the fundamental property of the gauge and clarifies conditions on gaugeability. For the proof, we refer to [62] (see § 5.6 and Theorem 4.19), or [33, Theorem 4.2], where it can be seen that the result is analogous to the Harnack inequality.

**Theorem 2.9 (Gauge Theorem).** *Let $D$ be a domain with $m(D) < \infty$ and let $q \in \mathcal{J}^\alpha$. If $u(x_0) < \infty$ for some $x_0 \in D$, then $u$ is bounded in $\mathbf{R}^d$. Moreover, the following conditions are equivalent:*

*(i) $(D, q)$ is gaugeable;*
*(ii) The semigroup $T_t$ satisfies $\int_0^\infty \|T_t\|_\infty \, dt < \infty$;*
*(iii) $V1 \in L^\infty(\mathbf{R}^d)$;*
*(iv) $V|q| \in L^\infty(\mathbf{R}^d)$.*

Thus, for the sake of brevity, we can write $V1 \in L^\infty(\mathbf{R}^d)$ to indicate that $(D, q)$ is gaugeable. In what follows we always assume that $(D, q)$ is gaugeable indeed. We like to remark that gaugeability is difficult to express explicitly. However a useful connection exists of gaugeability to the existence of positive functions harmonic on $D$ for $\Delta^{\alpha/2} + q$, which can be used to give natural and simple examples of the gauge function for some (not-so-natural) potentials $q$, see Figure 2.4 and [33].

The following estimate for the kernel $u_t(x, y)$ of the Feynman-Kac strengthens Lemma 4.7 in [32] and enables us to simplify the proof of **CGT**, compared to [32].

**Theorem 2.10.** *Let $D \subseteq \mathbf{R}^d$ be open with finite Lebesgue measure and $q \in \mathcal{J}_\alpha$. If $(D, q)$ is gaugeable and $0 < \delta < 1$ then for $x, y \in \mathbf{R}^d$*

$$u_t(x, y) \leqslant C_1 \, t^{-d/\alpha} \left[ (t^{-1/\alpha} |x - y|)^{-d-\alpha} \wedge 1 \right]^\delta \quad \text{for} \quad 0 < t \leqslant t_0 \, , \qquad (2.36)$$

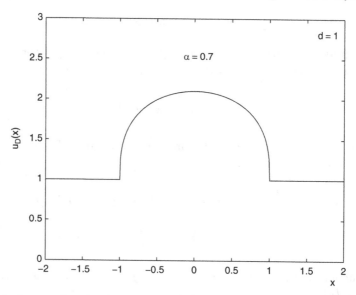

**Fig. 2.4** A gauge function for $q \geqslant 0$, see [33]

*where $t_0 = t_0(\delta, q, \alpha)$, $C_1 = C_1(d, \alpha)$ and*

$$u_t(x, y) \leqslant C_2 \exp(-\beta t), \quad for \quad t > t_0, \tag{2.37}$$

*where $C_2 = C_2(D, d, \alpha)$, $\beta = \beta(\delta, q, \alpha)$. Furthermore, if $D$ is additionally bounded then*

$$V(x, y) \leqslant C_3 |x - y|^{\alpha - d}. \tag{2.38}$$

*Proof.* Let $p > 1$ be fixed. Choose $t_0 > 0$ such that for $0 < t \leqslant t_0$

$$\sup_{x \in D} E^x[e_{2pq}(t)] \leqslant 2.$$

For $0 < t \leqslant t_0$ and $f \in L^2(D)$ define $S_t f(x) = E^x[t < \tau_D;\, e_{pq}(t)\, f(Y_t)]$. By Schwarz inequality, for $0 < t < t_0$ we have

$$|S_t f(x)|^2 \leqslant E^x[t < \tau_D;\, e_{2pq}(t)]\, E^x[t < \tau_D;\, f(Y_t)^2] \leqslant 2\, E^x[f(Y_t)^2]$$
$$= 2 \int f(y)^2\, p_t(x, y)\, dy \leqslant 2\, t^{-d/\alpha} \|f\|_2^2 \sup_x p_1(x).$$

Thus, we have obtained

$$\|S_t\|_{2,\infty} \leqslant C\, t^{-d/2\alpha}.$$

Now, observe that for positive $f \in L^1$ and positive $\phi \in L^2$ we have

$$\int \phi \, S_t f \, dx = \int f \, S_t \phi \, dx \leqslant ||S_t \phi||_\infty \int f \, dm \leqslant ||S_t||_{2,\infty} \, ||\phi||_2 \, ||f||_1 .$$

This shows that $S_t f \in L^2$ and $||S_t||_{1,2} \leqslant ||S_t||_{2,\infty}$. If $f \in L^1$ we have $S_t f = S_{t/2} S_{t/2} f \in L^\infty$ so

$$||S_t||_{1,\infty} \leqslant ||S_{t/2}||_{1,2} \, ||S_{t/2}||_{2,\infty} \leqslant ||S_{t/2}||_{2,\infty}^2 \leqslant C \, t^{-d/\alpha} .$$

Let $B$ be a Borel subset of $D$. Then

$$\begin{aligned}
T_t \mathbf{1}_B(x) &\leqslant E^x [t < \tau_D; \, e_{pq}(t) \, \mathbf{1}_B(Y_t)]^{1-\delta} \, E^x [t < \tau_D; \, \mathbf{1}_B(Y_t)]^\delta \\
&= (S_t \mathbf{1}_B(x))^{1-\delta} \, \left( P_t^D \mathbf{1}_B(x) \right)^\delta \\
&\leqslant ||S_t||_{1,\infty}^{1-\delta} \, m(B)^{1-\delta} \, [C_1 \, t^{-d/\alpha} ((t^{-1/\alpha} \rho)^{-d-\alpha} \wedge 1) \, m(B)]^\delta \\
&\leqslant C \, t^{-(1-\delta)d/\alpha} \, m(B) \, [C_1 \, t^{-d/\alpha} ((t^{-1/\alpha} \rho)^{-d-\alpha} \wedge 1)]^\delta \\
&\leqslant C_3 \, t^{-d/\alpha} \, m(B) \, [(t^{-1/\alpha} \rho)^{-d-\alpha} \wedge 1]^\delta ,
\end{aligned}$$

where $\rho = \sup_{y \in B} |x - y|$ and we applied the fact that

$$\begin{aligned}
p_t^D(x,y) &\leqslant p_t(x,y) = t^{-d/\alpha} \, p_1(t^{-1/\alpha}(x-y)) \\
&\leqslant C_1 \, t^{-d/\alpha} \, [(t^{-1/\alpha}|x-y|)^{-d-\alpha} \wedge 1], \quad t > 0, \, x, y \in \mathbb{R}^d .
\end{aligned}$$

Thus, we have obtained

$$\frac{1}{m(B)} \int_B u_t(x,y) \, dy \leqslant C_3 \, t^{-d/\alpha} \, m(B) \, [(t^{-1/\alpha} \rho)^{-d-\alpha} \wedge 1]^\delta .$$

Consider $B = B(y_0, \delta)$ and $x, y_0 \in D$. Letting $\delta \downarrow 0$, we obtain

$$u_t(x,y) \leqslant C_3 \, t^{-d/\alpha} \, [(t^{-1/\alpha} \, |x-y|)^{-d-\alpha} \wedge 1]^\delta ,$$

which gives (2.36). Since $(D,q)$ is gaugeable, by Theorem 2.5 we obtain

$$||T_t||_1 \leqslant C \, e^{-\epsilon t} ,$$

for $t > t_0$ and for some $\epsilon > 0$. On the other hand

$$\frac{1}{m(B)} \int u_t(x,y) \, \mathbf{1}_B(y) \, dy = ||T_t||_{1,\infty} .$$

As before, this gives $u_t(x,y) \leqslant ||T_t||_{1,\infty}$ . Since

$$||T_t||_{1,\infty} \leqslant ||T_{t_0}||_{1,\infty} \, ||T_{t-t_0}||_1 \leqslant ||T_{t_0}||_{1,\infty} \, C \, e^{-\epsilon(t-t_0)} ,$$

we obtain (2.37). To prove (2.38) we apply (2.36) to estimate

$$C_1^{-1} \int_0^{t_0} u_t(x,y)\, dt \leqslant \int_0^{t_0} t^{-d/\alpha}[1 \wedge (t^{-1/\alpha}|x-y|)^{-d-\alpha}]^\delta\, dt$$

$$= \int_0^{t_0} t^{-d/\alpha}[1 \wedge \left(\frac{t}{|x-y|^\alpha}\right)^{-d-\alpha}]^\delta\, dt$$

$$= |x-y|^{-\delta(d+\alpha)} \int_0^{|x-y|^\alpha \wedge t_0} t^{\delta-(1-\delta)d/\alpha}\, dt$$

$$+ \int_{|x-y|^\alpha \wedge t_0}^{t_0} t^{-d/\alpha}\, dt.$$

In the first integral on the right hand-side we take a $\delta > 0$ such that $d/\alpha < \frac{1+\delta}{1-\delta}$, and we then see that the first integral is convergent. We obtain the upper bound

$$C_4|x-y|^{-\delta(d+\alpha)}|x-y|^{(1-d/\alpha+\delta(1+d/\alpha))\alpha} + C_5|x-y|^{\alpha-d} = C\,|x-y|^{\alpha-d}.$$

To finish the proof we observe that (2.37) yields

$$\int_{t_0}^\infty u_t(x,y)\, dt \leqslant \beta^{-1}e^{-\beta t_0},$$

which, together with the observation that $|x-y| \leqslant \operatorname{diam}(D)$, concludes the proof of (2.38). □

We note that since $u_t(x,y)$ is continuous, the above estimate yields the continuity of $V(x,y)$ for $x,y \in \mathbb{R}^d$, $x \neq y$, under the assumption that $D$ is regular (and bounded).

We should also mention that there exists a new method of estimating $u_t$ based on the notion of *conditional smallness* of $q$ which yields *comparability* of $u_t$ and $p_t$ in finite time, see [36].

The following lemma is a well-known but fundamental relationship between $G_D$ and $V$, see [62], Ch. 6. For an analyst, the lemma is an instance of the (implicit) perturbation formula for $V$, compare [85].

**Lemma 2.11.** *Suppose that $q \in \mathcal{J}_\alpha$ and $V1 \in L^\infty(D)$. If $V|q|G_D|f| < \infty$ on $D$ then*

$$Vf = G_D f + VqG_D f.$$

*Proof.* By Fubini's theorem we obtain

$$VqG_D f(x) = E^x \int_0^{\tau_D} e_q(t)\, q(Y_t)\, E^{Y_t}[\int_0^{\tau_D} f(Y_s)\, ds]\, dt$$

$$= E^x \int_0^{\tau_D} e_q(t)\, q(Y_t) \int_t^{\tau_D} f(Y_s)\, ds\, dt$$

$$= E^x \int_0^{\tau_D} f(Y_s) \int_s^{\tau_D} e_q(t) \, q(Y_t) \, dt \, ds$$

$$= E^x \int_0^{\tau_D} f(Y_s) \left[ e_q(s) - 1 \right] ds = V \, f(x) - G_D f(x).$$

The application of Fubini's theorem is justified by the condition $V \, |q| \, G_D \, |f| < \infty$. □

An important consequence of the above lemma and Theorem 2.10 is the following

**Lemma 2.12.** *Suppose that* $q \in \mathcal{J}_\alpha$ *and* $V1 \in L^\infty(D)$. *Assume that* $D$ *is bounded and regular. Then for every* $x, y \in D$, $x \neq y$ *we have*

$$V(x, y) = G_D(x, y) + \int_D V(x, w) \, q(w) \, G_D(w, y) \, dw. \qquad (2.39)$$

*Proof.* Applying the preceding lemma we obtain that for every $x \in D$ the equation (2.39) holds $y$-almost everywhere. Assume that $|x - y| > \delta > 0$. Then either $|x - w| > \delta/2$ or $|w - y| > \delta/2$. Suppose that the first condition holds. Then, by Theorem 2.10 we obtain $V(x, w) \leqslant C \, K_\alpha(x, w) \leqslant C \, K_\alpha(\delta/2)$ so

$$V(x, w) \, |q(w)| \, G_D(w, y) \leqslant C \, K_\alpha(\delta/2) \, |q(w)| \, K_\alpha(w - y).$$

In the second case we obtain

$$V(x, w) \, |q(w)| \, G_D(w, y) \leqslant C \, K_\alpha(x - w) \, |q(w)| \, K_\alpha(\delta/2).$$

Consequently, when $|x - y| > \delta$ we have

$$V(x, w) \, |q(w)| \, G_D(w, y) \leqslant C \, K_\alpha(\delta/2) \, |q(w)| \, [K_\alpha(w - y) + K_\alpha(x - w)].$$

Since $D$ is bounded, it follows that the set of functions

$$\{w \mapsto V(x, w) \, q(w) \, G_D(w, y); (x, y) \in D \times D, |x - y| > \delta\}$$

is uniformly integrable on $D$. On the other hand, for each $w \in D$, the function

$$(x, y) \mapsto V(x, w) \, q(w) \, G_D(w, y)$$

is continuous except possibly at $x = w$ or $y = w$. Therefore, the integral on the right-hand side of (2.39) is continuous in $(x, y) \in D \times D$, $|x - y| > \delta$. Since $\delta$ is arbitrary, both members of (2.39) are continuous in $(x, y) \in D \times D$, $x \neq y$. The proof is complete. □

Let $h$ be an $\alpha$-harmonic and positive on a bounded domain $D$. By $p_t^D(x, y)$ we denote the transition density function of $(Y_t)$ killed on exiting $D$. For $x, y \in D$ and $t > 0$ we define time-homogeneous transition density (of Doob's $h$-process)

$$p_h(t; x, y) = h(x)^{-1} p_t^D(x, y) h(y).$$

This defines a strong Markov process on $D_\partial = D \cup \{\partial\}$. Here $\partial$ is the absorbing state (cemetery) attached to the state space to accommodate for the loss of mass (the conditional process is generally subprobabilistic if considered on the original state space, [23]). The $h$-process is denoted also by $Y_t$, while the corresponding expectations and probabilities are denoted by $E_h^x, P_h^x$. We should note that even though we use the same generic notation for the conditional process, there is no pathwise correspondence between the original and the conditional processes, and theorems involving the conditional process are usually more difficult.

The definition of the $h$-process yields,

$$E_h^x[t < \tau_D; f(Y_t)] = h(x)^{-1} E^x[t < \tau_D; f(Y_t) h(Y_t)]. \qquad (2.40)$$

Let $D$ be a bounded Lipschitz domain; for fixed $\xi \in \partial D$ we put

$$h(y) = M_D(y, \xi).$$

Here $M_D(\cdot, \xi)$ is Martin's kernel of $D$, which is $\alpha$-harmonic in $D$ [27, 38].

We also need another version of conditioning: for fixed $y \in D$ we let

$$h(y) = G_D(y, z).$$

The function $h$ above is $\alpha$-harmonic in $D \setminus \{z\}$, and superharmonic in $D$, see [56]. In the sequel, we will use the notation $E_\xi^x, P_\xi^x$ ($E_z^x, P_z^x$, respectively) to indicate conditioning by Martin kernel (Green function, respectively).

We redefine $\alpha$-stable $\xi$-Lévy motion $Y_t$ by putting $Y_s = \xi$ for $s \geqslant \tau_D$ to get the process on $D \cup \{\xi\}$. Analogously, $\alpha$-stable $z$-Lévy motion $Y_t$ is defined on $D$ and $Y_s = z$ for $s \geqslant \tau_{D \setminus \{z\}}$.

For a stopping time $T \leqslant \tau_{D \setminus \{y\}}$ we obtain a specialization of the formula (2.40):

$$E_z^x[T < \tau_{D \setminus \{y\}}; f(Y_T)] = G_D(x, z)^{-1} E^x[T < \tau_D; f(Y_T) G_D(Y_T, z))]. \quad (2.41)$$

A similar formula holds true for the $\xi$-process.

As an instructive exercise we compute the Green function of $\alpha$-stable $z$-Lévy motion.

**Proposition 2.13 (Green function for conditional process).** *Let* $D$ *be a bounded Lipschitz domain in* $\mathbb{R}^d$, $\alpha < d$, $Y_t$ - $\alpha$-stable $z$-Lévy motion. *The Green function of* $D$, *as the function of* $y$, *computed for the process* $Y$

*(starting at $x \in D$ and exiting at $z \in D$) has the following form:*

$$\frac{G_D(x,y)\,G_D(y,z)}{G_D(x,y)}.$$

*Proof.* Indeed, we obtain

$$
\begin{aligned}
E_z^x \int_0^{\tau_{D\setminus\{z\}}} f(Y_t)\,dt &= \int_0^\infty E_z^x[t < \tau_{D\setminus\{z\}}; f(Y_t)]\,dt \\
&= \int_0^\infty G_D(x,z)^{-1}E^x[t < \tau_{D\setminus\{z\}}; f(Y_t)G_D(Y_t,z)]\,dt \\
&= G_D(x,z)^{-1}E^x \int_0^{\tau_D} f(Y_t)\,G_D(Y_t,z)\,dt \\
&= G_D(x,z)^{-1}G_D(f(\cdot)\,G_D(\cdot,z))(x) \\
&= \int_D \frac{G_D(x,y)\,G_D(y,z)}{G_D(x,y)}\,f(y)\,dy\,.
\end{aligned}
$$

$\square$

By the above calculations, we obtain

$$
E_z^x \tau_{D\setminus\{z\}} = E_z^x \int_0^{\tau_{D\setminus\{z\}}} 1(Y_t)\,dt = \int_D \frac{G_D(x,y)\,G_D(y,z)}{G_D(x,y)}\,dy
$$

$$
\leqslant C \int_D [K_\alpha(x-y) + K_\alpha(y-z)]\,dy \leqslant 2C \int_{D\cap B(0,R)} K_\alpha(y)\,dy < \infty\,.
$$

The calculations provide the proof of the second formula in Theorem 2.14 below (the proof of the first formula is similar and will be omitted).

**Theorem 2.14.**

$$
E_\xi^x \tau_D < \infty\,, \qquad P_\xi^x(\lim_{t\uparrow\tau_D} Y_t = \xi) = 1\,,
$$

$$
E_z^x \tau_{D\setminus\{z\}} < \infty\,, \qquad P_z^x(\lim_{t\uparrow\tau_{D\setminus\{z\}}} Y_t = z) = 1\,.
$$

Theorem 2.14 shows that the behavior of the conditional process is dramatically different from that of the original process. In particular, the conditional process exits through the pole of the function $h$ and does so in a continuous manner.

**Lemma 2.15.** *Let $D, U$ be bounded regular (e.g. Lipschitz) domains such that $\overline{U} \subseteq D$ and $z \in U$. Put $D_0 = D \setminus \overline{U}$ and let $\zeta = \tau_{D\setminus\{z\}}$. Let $u \in D$, $u \neq z$ and $x \in D_0$. Then we have*

$$
P_z^u\{\tau_{U\setminus\{z\}} = \zeta\} = \frac{G_U(u,z)}{G_D(u,z)}\,, \tag{2.42}
$$

$$
P_z^x\{\tau_{D_0} = \zeta\} = 0\,. \tag{2.43}
$$

Let us remark that the second formula states that the conditional process cannot reach the point $z$ jumping from outside a certain neighborhood of this point. The first formula gives the precise value of the probability of reaching the point $z$ when starting from a point within a neighborhood of this point. The formulas are essential in proving **CGT** for the operator $\Delta^{\alpha/2}$, $0 < \alpha < 2$.

*Proof.* To prove the first formula we observe that

$$G_D(u,v) = K_\alpha(u - v) - E^u[K_\alpha(Y_{\tau_D} - v)].$$

Consequently, we obtain

$$
\begin{aligned}
E^u[\tau_U < \tau_D; G_D(Y_{\tau_U}, z)] &= E^u[\tau_U < \tau_D; K_\alpha(Y_{\tau_U} - z)] \\
&\quad - E^u[\tau_U < \tau_D; E^{Y_{\tau_U}}[K_\alpha(Y_{\tau_D} - z)]] \\
&= E^u[\tau_U < \tau_D; K_\alpha(Y_{\tau_U} - z)] \\
&\quad - E^u[\tau_U < \tau_D; E^u[K_\alpha(Y_{\tau_D} - z) \circ \theta_{\tau_U} | \mathcal{F}_{\tau_U}]] \\
&= E^u[\tau_U < \tau_D; K_\alpha(Y_{\tau_U} - z) - K_\alpha(Y_{\tau_D} - z)] \\
&= E^u[K_\alpha(Y_{\tau_U} - z) - K_\alpha(Y_{\tau_D} - z)] \\
&= G_D(u,z) - G_U(u,z).
\end{aligned}
$$

Taking into account

$$P_z^u\{\tau_{U\setminus\{z\}} \neq \zeta\} = G_D(u,z)^{-1} E^u[\tau_U < \tau_D; G_D(Y_{\tau_U}, z)],$$

we obtain the first formula. To prove the second formula we observe that $G_D(\cdot, z)$ is $\alpha$-harmonic and bounded on $D_0 = D \setminus \overline{U}$ so

$$E^x G_D(Y_{\tau_{D_0}}, z) = G_D(x,z)$$

and

$$
\begin{aligned}
P_z^x\{\tau_{D_0} < \zeta\} &= G_D(x,z)^{-1} E^x[\tau_{D_0} < \tau_D; G_D(Y_{\tau_{D_0}}, z)] \\
&= G_D(x,z)^{-1} E^x G_D(Y_{\tau_{D_0}}, z) = G_D(x,z)^{-1} G_D(x,z) = 1.
\end{aligned}
$$

$\square$

As a corollary we obtain (compare Lemma 4.4 in [64]):

**Corollary 2.16.** *Assume that* $y \in D$ *with* $d(y, D^c) > 3\delta$. *Put* $U = B(y, 3\delta)$. *Then we have*

$$\inf_{u \in \overline{B(y,\delta)}} P^u\{\tau_U = \tau_D\} > 0, \qquad \inf_{u \in \overline{B(y,\delta)}\setminus\{y\}} P_y^u\{\tau_{U\setminus\{y\}} = \zeta\} > 0. \qquad (2.44)$$

*Proof.* We first prove the second part of (2.44). In view of Lemma 2.15 we have that

$$P_y^u\{\tau_{U\setminus\{y\}} = \zeta\} = \frac{G_U(u,y)}{G_D(u,y)} \geqslant \frac{G_U(u,y)}{K_\alpha(u,y)} = 1 - \mathcal{A}_{d,\alpha}^{-1} |u-y|^{d-\alpha} E^u K_\alpha(Y_{\tau_U}, y).$$

Observe that we have $|u-y| \leqslant \delta$ for $u \in \overline{B(y,\delta)}$ and also $|Y_{\tau_U} - y| > 3\delta$ which yields $E^u K_\alpha(Y_{\tau_U}, y) \leqslant \mathcal{A}_{d,\alpha}(3\delta)^{\alpha-d}$. This completes the proof of the second part of (2.44). We now prove the first part. We denote $R = \operatorname{diam}(D)$. Then, by the explicit formula for Poisson kernel for balls (1.57), we obtain

$$
\begin{aligned}
P^u\{\tau_U = \tau_D\} &= P^u\{Y_{\tau_U} \in D^c\} \\
&= P^{u-y}\{Y_{\tau_{B(0,3\delta)}} \in D^c - y\} \geqslant P^{u-y}\{Y_{\tau_{B(0,3\delta)}} \in B(0,R)^c\} \\
&= C_\alpha^d \int_{|z|>R} \left( \frac{(3\delta)^2 - |u-y|^2}{|z|^2 - (3\delta)^2} \right)^{\alpha/2} \frac{dz}{|u-y-z|^d} \\
&\geqslant C_\alpha^d (8\delta)^{\alpha/2} \omega_d \int_R^\infty \frac{\rho^{d-1}\, d\rho}{\rho^\alpha\, (\rho+\delta)^d} \geqslant \frac{C_\alpha^d\, (8\delta)^{\alpha/2} \omega_d}{(7/6)^d\, \alpha\, R^\alpha},
\end{aligned}
$$

because we have $\rho \geqslant R > 6\delta$ under the integral sign. By $\omega_d$ we denote the surface measure of the unit sphere in $\mathbf{R}^d$.                                     □

The following lemma is a "conditional" version of Khasminski's lemma (see Lemma 2.8). The proof relies on **3G** Theorem (2.26) as in [64].

**Lemma 2.17.** *For every $\varepsilon > 0$ there exists $\eta = \eta(\varepsilon, D, q)$ such that for every open set $U \subseteq D$ with $m(U) < \eta$ we have*

$$\sup_{u \in D, u \neq y} E_y^u \int_0^{\tau_{U\setminus\{y\}}} |q(Y_t)|\, dt < \varepsilon$$

*and if $0 < \varepsilon < 1$ then $\exp(-\varepsilon) \leqslant E_y^u e_q(\zeta) \leqslant (1-\varepsilon)^{-1}$.*

*Proof.* Let $x, y$ be in $D$, $x \neq y$. Applying the definition of transition probability $p_h^D$ of the process conditioned by the function $h(\cdot) = G(\cdot, y)$ and using Fubini's Theorem we obtain

$$
\begin{aligned}
E_y^x\left[ \int_0^{\tau_U} |q(Y_t)|\, dt \right] &\leqslant E_y^x\left[ \int_0^{\tau_D} 1_U(Y_t) |q(Y_t)|\, dt \right] \\
&= \int_0^\infty \int_U p_{G(\cdot,y)}^D(t; x, u)\, |q(u)|\, du\, dt \\
&= G(x,y)^{-1} \int_0^\infty \int_U p^D(t; x, u)\, |q(u)|\, G(u,y)\, du\, dt \\
&= G(x,y)^{-1} \int_U G(x,u)\, |q(u)|\, G(u,y)\, du.
\end{aligned}
$$

By **3G** Theorem, the last integral is estimated by

$$C \int_U [K_\alpha(x,u) + K_\alpha(u,y)]\, |q(u)|\, du,$$

with $C$ depending only on $D$ and $q$. However, by the properties of Kato class, we obtain

$$\sup_{x \in D} \int_U K_\alpha(x, u) \, |q(u)| \, du \longrightarrow 0 \, ,$$

as $m(U) \longrightarrow 0$. The last part of the lemma now follows from Khasminski's Lemma. ☐

By the above lemma, Lemma 2.8 and Corollary 2.44, we easily obtain the following result (compare Lemma 4.3 in [64]).

**Lemma 2.18.** *Under the notation of Corollary 4.4 there exist constants $C_1$ and $C_2$ such that for every $u, v \in \overline{B(y, \delta)}$, $v \neq y$, with $\delta > 0$ small enough we have*

$$C_1 \leqslant E^u[\tau_U = \tau_D; \, e_q(\tau_D)] \leqslant C_2 \, , \quad C_1 \leqslant E_y^v[\tau_{U \setminus \{y\}} = \zeta; \, e_q(\zeta)] \leqslant C_2 \, .$$

$$(2.45)$$

*Proof.* We prove the second part of (2.18). The other case is similar and is left to the reader. Applying Lemma 2.17 with $\varepsilon = 1/2$, we obtain $E_y^u[e_{|q|}(\tau_U)] \leqslant 2$. Denote the infimum in Corollary 2.16 by $C$. Then, in view of Jensen's inequality, we obtain

$$\begin{aligned}
E_y^u[e_q(\tau_U)|\tau_U = \zeta] &\geqslant \exp\{E_y^u[\int_0^{\tau_U} q(Y_t)\,dt|\tau_U = \zeta]\} \\
&\geqslant \exp\{-E_y^u[\int_0^{\tau_U} |q(Y_t)|\,dt|\tau_U = \zeta]\} \\
&\geqslant \exp\{-\frac{1}{C} E_y^u[\tau_U = \zeta; \int_0^{\tau_U} |q(Y_t)|\,dt]\} \\
&\geqslant \exp\{-\frac{1}{C} E_y^u[\int_0^{\tau_U} |q(Y_t)|\,dt]\} \geqslant \exp\{-\frac{1}{2C}\} \, .
\end{aligned}$$

Before stating the next lemma, we introduce some notation. For $y \in \mathbf{R}^d$, $|y| > 1$ let

$$I_1(y) = \int_{B(0,1)} \frac{\mathcal{A}_{d,\alpha}\,du}{|u - y|^{d+\alpha}|u|^{d-\alpha}} \, , \quad I_2(y) = \int_{B(0,1)} \frac{\mathcal{A}_{d,\alpha}\,du}{|u - y|^{d+\alpha}} \, . \quad (2.46)$$

**Lemma 2.19.** *For all $y \in \mathbf{R}^d$ such that $|y| > 1$ we have*

$$I_1(y) \approx I_2(y) \, .$$

*Proof.* Clearly, we have $I_1(y) \geqslant I_2(y)$. To show a reverse inequality, denote $A(y) = B(y/|y|, 1/2) \cap B(0,1)$, $B(y) = B(0,1) \setminus A(y)$ and $M(y) = \sup_{u \in B(y)} |y - u|^{-d-\alpha}$, $m(y) = \inf_{u \in B(y)} |y - u|^{-d-\alpha}$. It is not difficult to see that $M(y) \leqslant Cm(y)$. Consequently,

$$\int_{B(y)} \frac{du}{|u-y|^{d+\alpha}|u|^{d-\alpha}} \leqslant M(y) \int_{|u|<1} \frac{du}{|u|^{d-\alpha}} \leqslant C \frac{\omega_d}{\alpha} m(y)$$

$$\leqslant C \frac{\omega_d}{\alpha |B(y)|} \int_{B(y)} \frac{du}{|u-y|^{d+\alpha}}.$$

However, for $u \in A(y)$ we have $|u| > 1/2$, so

$$\int_{A(y)} \frac{du}{|u-y|^{d+\alpha}|u|^{d-\alpha}} \leqslant 2^{d-\alpha} \int_{A(y)} \frac{du}{|u-y|^{d+\alpha}}.$$

$\square$

We define the *conditional gauge* as the gauge function for the conditional process:

$$u(x,y) = E_y^x e_q(\tau_{D \setminus \{y\}}), \quad x \in D, \quad y \in \overline{D}.$$

Recall the Ikeda-Watanabe formula (1.52): for a bounded domain $D$ with the exterior cone property the density function of the $P^x$-distribution of $Y_{\tau_D}$ is given, for $x \in D$, by

$$\mathcal{A}_{d,-\alpha} \int_D \frac{G_D(x,v)}{|v-y|^{d+\alpha}} \, dv, \quad y \in D^c.$$

The following explains the role of the conditional gauge function (compare [62, Theorem 6.3]).

**Lemma 2.20.** *If $(D, q)$ is gaugeable then*

$$V(x,y) = u(x,y) \, G_D(x,y), \quad x,y \in D, \quad x \neq y. \tag{2.47}$$

*Proof.* Since $x \neq y$, by the proof of Lemma 2.12 we obtain that

$$V(x, \cdot) \, |q|(\cdot) \, G_D(\cdot, y) < \infty$$

on the set $\{(x,y) \in D \times D, |x-y| > \delta\}$, for a fixed $\delta > 0$. Applying Fubini's theorem,

$$E_y^x \int_0^\zeta e_q(t) \, q(Y_t) \, dt = \int_0^\infty E_y^x [t < \zeta; e_q(t) \, q(Y_t)] \, dt$$

$$= G_D(x,y)^{-1} \int_0^\infty E^x [t < \tau_D; e_q(t) \, q(Y_t) \, G_D(Y_t, y)] \, dt$$

$$= G_D(x,y)^{-1} \int_D V(x,w) \, q(w) \, G_D(w,y) \, dw.$$

Since

$$E_y^x \int_0^\zeta e_q(t) \, q(Y_t) \, dt = E_y^x \int_0^\zeta \frac{d \, e_q(t)}{dt} \, dt = E_y^x [e_q(\zeta) - 1] = u(x,y) - 1,$$

by Lemma 2.12 we obtain (2.47). $\square$

The preceding lemma yields the following.

**Lemma 2.21.** *Let $D$ be an open regular bounded subset of $\mathbb{R}^d$. Then the gauge function $u(x,y)$ is continuous and symmetric on $D \times D$, $x \neq y$.*

We thus arrive at the main conclusion of this section: Conditional Gauge Theorem (**CGT**).

**Theorem 2.22 (CGT).** *Let $D$ be a bounded Lipschitz domain, $q \in \mathcal{J}^\alpha$. If $(D,q)$ is gaugeable (i.e. $E^x e_q(\tau_D) < \infty$) then*

$$\sup_{x,y \in D} u(x,y) < \infty,$$

*and, moreover, $u$ has a symmetric continuous extension to $\overline{D} \times \overline{D}$.*

*Proof.* The proof is carried out in several steps.

*Step 1.* For $\delta > 0$ we put $D_\delta = \{x \in D; d(x, D^c) > 3\delta\}$. We choose and fix throughout the proof $\delta$ and a Lipschitz domain $U^\delta$ such that $D \backslash D_\delta \subseteq U^\delta \subseteq D$ and for all $y \in D$

$$\sup_{u \in D, u \neq y} E_y^u \Big[ \int_0^\tau |q(Y_t)| \, dt \Big] < 1/2, \quad \sup_{u \in \mathbb{R}^d} E^u \Big[ \int_0^\tau |q(Y_t)| \, dt \Big] < 1/2,$$

with $\tau = \tau_{U^\delta \backslash \{y\}}$ or $\tau = \tau_{B(y,3\delta) \backslash \{y\}}$. By Lemma 2.8 and Lemma 2.17 we obtain

$$\sup_{u \in D, u \neq y} E_y^u e_{|q|}(\tau) \leqslant 2, \quad \sup_{u \in \mathbb{R}^d} E^u e_{|q|}(\tau) \leqslant 2.$$

We show for $x, y \in D_\delta$, $x \neq y$ the following:

$$u(x,y) < C, \quad \text{where } C = C(D, \alpha, q, \delta).$$

Fix $x, y \in D_\delta$ and denote $D_0 = D \backslash \overline{B(y,\delta)}$, $U = B(y,3\delta) \backslash \{y\}$ and

$$T_1 = \tau_{D_0}, \quad T_n = S_{n-1} + \tau_{D_0} \circ \theta_{S_{n-1}},$$
$$S_0 = 0, \quad S_n = T_n + \tau_U \circ \theta_{T_n}, \quad n = 1, 2, \ldots$$

Put $\zeta = \tau_{D \backslash \{y\}}$. Because of the second formula in Lemma 2.15, the (conditional) process exits from $D \backslash \{y\}$ first entering $B(y,\delta)$. Hence $\zeta = S_n$, for some $n$. Thus, we obtain

$$E_y^x e_q(\zeta) = \sum_{n=1}^\infty E_y^x [T_n < \zeta, S_n = \zeta; e_q(\zeta)]. \tag{2.48}$$

For $n = 1$, by strong Markov property, we obtain

$$G(x,y) E_y^x[T_1 < \zeta, S_1 = \zeta; e_q(\zeta)]$$
$$= G(x,y) E_y^x[\tau_{D_0} < \zeta; e_q(T_1)\{\tau_U = \zeta; e_q(\tau_U)\} \circ \theta_{T_1}]$$
$$= G(x,y) E_y^x[\tau_{D_0} < \zeta; e_q(T_1) E_y^{Y_{T_1}}[\tau_U = \zeta; e_q(\tau_U)]]$$
$$= E^x[\tau_{D_0} < \tau_D; e_q(T_1) G(Y_{T_1}, y) E_y^{Y_{T_1}}[\tau_U = \zeta; e_q(\tau_U)]].$$

Since $\tau_{D_0} < \tau_D$ yields $Y_{T_1} \in \overline{B(y,\delta)}$, using Lemma 2.44 we obtain that the last term above is equivalent to

$$E^x[\tau_{D_0} < \tau_D; e_q(T_1) G(Y_{T_1}, y)] \leqslant E^x[e_q(T_1) G(Y_{T_1}, y)].$$

Taking into account one term of the series (2.48), we get

$$G(x,y) E_y^x[T_n < \zeta, S_n = \zeta; e_q(\zeta)]$$
$$= G(x,y) E_y^x[T_n < \zeta; e_q(T_n)\{\tau_U = \zeta; e_q(\tau_U)\} \circ \theta_{T_n}]$$
$$= G(x,y) E_y^x[T_n < \zeta; e_q(T_n) E_y^{Y_{T_n}}[\tau_U = \zeta; e_q(\zeta)]]$$
$$= E^x[T_n < \tau_D; e_q(T_n) G(Y_{T_n}, y) E_y^{Y_{T_n}}[\tau_U = \zeta; e_q(\zeta)]].$$

Since $Y_{T_n} \in \overline{B(y,\delta)}$, whenever $T_n < \tau_D$, so by Lemma 4.5 we obtain

$$E^x[T_n < \tau_D; e_q(T_n) G(Y_{T_n}, y) E_y^{Y_{T_n}}[\tau_U = \zeta; e_q(\zeta)]]$$
$$\approx E^x[T_n < \tau_D; e_q(T_n) G(Y_{T_n}, y)]$$
$$= E^x[S_{n-1} < \tau_D; e_q(S_{n-1})\{\tau_{D_0} < \tau_D; e_q(\tau_{D_0}) G(Y_{\tau_{D_0}}, y)\} \circ \theta_{S_{n-1}}]$$
$$= E^x[S_{n-1} < \tau_D; e_q(S_{n-1}) E^{Y_{S_{n-1}}}[\tau_{D_0} < \tau_D; e_q(\tau_{D_0}) G(Y_{\tau_{D_0}}, y)]]$$
$$= E^x[S_{n-1} < \tau_D; e_q(S_{n-1}) E^{Y_{S_{n-1}}}[e_q(\tau_{D_0}) G(Y_{\tau_{D_0}}, y)]].$$

Using Ikeda-Watanabe formula for $D_0$ and Lemma 2.19, we have for $z \in D_0$

$$E^z[e_q(\tau_{D_0}) G(Y_{\tau_{D_0}}, y)] = \int_{D_0} \int_{D_0^c \cap D} \widetilde{u}(z,v) G(w,y) \frac{\mathcal{A}_{d,-\alpha}}{|v-w|^{d+\alpha}} G_{D_0}(z,v) \, dw \, dv$$
$$\leqslant \int_{D_0} \int_{B(y,\delta)} \widetilde{u}(z,v) K_\alpha(w,y) \frac{\mathcal{A}_{d,-\alpha}}{|v-w|^{d+\alpha}} G_{D_0}(z,v) \, dw \, dv$$
$$= \int_{D_0} \widetilde{u}(z,v) \mathcal{A}_{d,-\alpha} \delta^{-d} I_1\left(\frac{y-v}{\delta}\right) G_{D_0}(z,v) \, dv$$
$$\approx \int_{D_0} \widetilde{u}(z,v) \mathcal{A}_{d,-\alpha} \delta^{\alpha-d} \delta^{-\alpha} I_2\left(\frac{y-v}{\delta}\right) G_{D_0}(z,v) \, dv$$
$$\approx \delta^{\alpha-d} \int_{D_0} \int_{B(y,\delta)} \widetilde{u}(z,v) \frac{\mathcal{A}_{d,-\alpha}}{|v-w|^{d+\alpha}} G_{D_0}(z,v) \, dw \, dv$$
$$\leqslant \delta^{\alpha-d} \int_{D_0} \int_{D_0^c} \widetilde{u}(z,v) \frac{\mathcal{A}_{d,-\alpha}}{|v-w|^{d+\alpha}} G_{D_0}(z,v) \, dw \, dv$$
$$= \delta^{\alpha-d} E^z e_q(\tau_{D_0}) \approx \delta^{\alpha-d} E^z e_q(\tau_D) \approx \delta^{\alpha-d};$$

by gaugeability. Here $\widetilde{u}(z,v) = \widetilde{E}_v^z e_q(\tau_{D_0 \setminus \{v\}})$ is the conditional gauge of the set $D_0$.

If $T_{n-1} < \tau_D$, for $n \geqslant 2$, then $Y_{T_{n-1}} \in \overline{B(y,\delta)}$ hence by (4) and Lemma 4.5 we obtain

$$
\begin{aligned}
G(x,y)E_y^x[T_n < \zeta, S_n = \zeta; e_q(\zeta)] &\leqslant C\delta^{\alpha-d}E^x[S_{n-1} < \tau_D; e_q(S_{n-1})] \\
&= C\delta^{\alpha-d}E^x[T_{n-1} < \tau_D; e_q(T_{n-1})\{\tau_U < \tau_D; e_q(\tau_U)\} \circ \theta_{T_{n-1}}] \\
&= C\delta^{\alpha-d}E^x[T_{n-1} < \tau_D; e_q(T_{n-1})E^{Y_{T_{n-1}}}[\tau_U < \tau_D; e_q(\tau_U)]] \\
&\leqslant 2C\delta^{\alpha-d}E^x[T_{n-1} < \tau_D; e_q(T_{n-1})] \\
&\approx 2C\delta^{\alpha-d}E^x[T_{n-1} < \tau_D; e_q(T_{n-1})E^{Y_{T_{n-1}}}[\tau_U = \tau_D; e_q(\tau_U)]] \\
&= 2C\delta^{\alpha-d}E^x[T_{n-1} < \tau_D, S_{n-1} = \tau_D; e_q(\tau_D)].
\end{aligned}
$$

Thus

$$
\begin{aligned}
G(x,y)\, E_y^x e_q(\zeta) = G(x,y) \sum_{n=1}^{\infty} E_y^x[T_n < \zeta, S_n = \zeta;\, e_q(\zeta)] \\
\leqslant C\delta^{\alpha-d}\left(1 + \sum_{n=2}^{\infty} E^x[T_{n-1} < \tau_D, S_{n-1} = \tau_D;\, e_q(\tau_D)]\right) \\
\leqslant C\delta^{\alpha-d}\left(1 + E^x e_q(\tau_D)\right).
\end{aligned}
$$

Recall that $x,y$ satisfy the conditions $d(x,D^c) > 3\delta$, $d(y,D^c) > 3\delta$ and $|x-y| \leqslant \operatorname{diam}(D) < \infty$. We obtain (compare [62], Lemma 6.7)

$$
G(x,y) \geqslant C'|x-y|^{\alpha-d} \geqslant C' \left(\operatorname{diam}(D)\right)^{\alpha-d},
$$

with $C' = C'(D,\alpha,q,\delta)$. This clearly ends the proof of Step 1.

*Step 2.* In this step we remove the condition $d(x,D^c) > 3\delta$ imposed on $x \in D$ in (5).

To do this, assume that $y \in D_\delta$ but $d(x,D^c) \leqslant 3\delta$. Let $U^\delta$ be as in Step 1. Denote $U = U^\delta \setminus \{y\}$. Then we have

$$
\begin{aligned}
u(x,y) &= E_y^x[\tau_U = \zeta;\, e_q(\tau_U)] + E_y^x[\tau_U < \zeta;\, e_q(\tau_U)\, u(Y_{\tau_U}, y)] \\
&\leqslant E_y^x e_{|q|}(\tau_U)\Big(1 + \sup_{w \in D_\delta, w \neq y} u(w,y)\Big).
\end{aligned}
$$

By Step 1 and properties of $U^\delta$, we obtain the conclusion.

*Step 3.* In this step we apply the symmetry of the function $u(x,y)$ in $x,y \in D$ to finish the proof of the boundedness of $u$.

Observe that the symmetry of $u$ along with Step 2 settle the case when $x \in D_\delta$ and $y \in D_\delta^c$. It remains only the case when $x \neq y$, $x,y \in D_\delta^c$.

To resolve this case, we proceed exactly as in Step 2 to obtain

$$u(x,y) = E_y^x[\tau_U = \zeta;\ e_q(\tau_U)] + E_y^x[\tau_U < \zeta;\ e_q(\tau_U)\, u(Y_{\tau_U}, y)].$$

If $\tau_U < \zeta$ then $d(Y_{\tau_U}, D^c) > 3\delta$ which reduces the proof to the case $x \in D_\delta$, $y \in D_\delta^c$. By Step 2 and symmetry of $u$, we obtain the conclusion. This completes the proof of the theorem.                                                          □

**Concluding remarks.** We like to note that in the proof of **CGT** for $\Delta$ ([62]) one first considers conditioning by the boundary (Martin kernel). The boundedness of the conditional gauge for interior points of the domain is then obtained as an easy corollary by considering the further evolution of the Brownian motion till it hits the boundary. For $\Delta^{\alpha/2}$ ($0 < \alpha < 2$), due to the jumps of the process, the more important is the boundedness of the conditional gauge in the interior of $D$ (conditioning by the Green function), and it cannot be obtained easily from the boundary behavior of the conditional gauge. Instead, we obtain the boundedness of the conditional gauge on the boundary as an easy corollary by approximation from within the domain.

We should also observe that the recent advances in the understanding of the role of the **3G** Theorem allow for *analytic* proofs of **CGT** in this and other settings, by using the perturbation series. We refer the reader to [85] for details, and to [84] for the general perspective on the role of **BHP** in proving **3G**. Such an approach has the advantage of being more explicit, algebraic, and *discrete*, paralleling the definition of the exponential function in terms of the power series, rather than differential equations. On the other hand, the probabilistic setting allows for intrinsic interpretations and verbalization of the proofs in terms of mass and trajectories of stochastic processes. The authors may only wonder which of these two approaches is more the reality, and which is more the language.

We want to conclude our discussion by mentioning a few directions of further research. First, it seems important to obtain an approximate factorization of the Green function for general (non-Lipschitz) domains, by using [38]. Second, it is of interest to study the asymptotics of the Martin kernel for narrow cones, and use the setup of [5] to complete the results of [111]. Third, it is of paramount importance to give sharp estimates for the *transition density of the killed process*. Fourth, it seems important to generalize the results discussed above to other stable Lévy processes ([40]), to more general jump type Markov processes, and to more general additive perturbations of their generators ([36, 52, 102, 82, 83]).

# Chapter 3
# Nontangential Convergence for $\alpha$-harmonic Functions

by M. Ryznar

## 3.1 Introduction

Let $D$ be the open unit circle in $\mathbb{R}^2$ and $f$ be a bounded classical harmonic function on $D$; that is $\Delta f = 0$ on $D$. In 1906 Fatou (see [75]) proved that

$$\lim_{x \to Q \in \partial D} f(x) \quad \text{exists a.s.,}$$

where the limit is nontangential. That is $x \to Q \in \partial D$ and $x \in \Gamma_Q$ - a bounded cone with vertex $Q$ which is included in $D$ (see picture below).

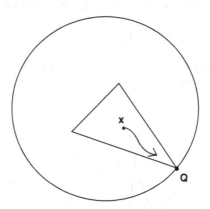

Littlewood in 1927 (see [120]) proved the sharpness of Fatou's result. For any tangential curve $\gamma$ contained in the disk $D$ ending at 1 there is a bounded harmonic function such that for almost every $\theta$ the limit

$$\lim_{x \to e^{i\theta}, x \in \gamma e^{i\theta}} f(x) \quad \text{does NOT exist.}$$

K. Bogdan et al., *Potential Analysis of Stable Processes and its Extensions*, Lecture Notes in Mathematics 1980, DOI 10.1007/978-3-642-02141-1_3, © Springer-Verlag Berlin Heidelberg 2009

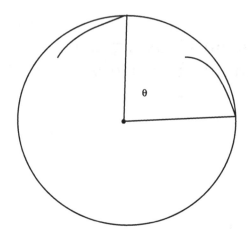

The theorem of Fatou was later extended to much more general domains in $\mathbb{R}^d$. The following authors contributed in solving the problem of the nontangential convergence under various assumptions on the domain $D$ using either analytical or probabilistic methods: Hunt and Wheeden (1968) [94], Ancona (1978) [3], Jerison and Kenig (1982) [100], Doob(1984) [71], Bass(1995) [10], Chen, Durret and Ma (1997) [46].

**Theorem 3.1 (Fatou's Theorem).** *If $f \geq 0$ is a harmonic function on a Lipschitz domain $D$, that is $\Delta f = 0$ on $D$, then the nontangential (finite) limit*

$$\lim_{x \to Q \in \partial D} f(x)$$

*exists a.s. with respect to the surface measure of $\partial D$. The limit has to be taken nontangentially, that is $|x - Q| < t\delta_x(D)$, where $t > 0$ is fixed.*

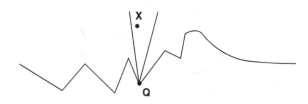

Another problem which may be of interest is the boundary behavior of ratios of positive harmonic functions. Does the nontangential limit $\lim_{x \to Q \in \partial D} \frac{f(x)}{g(x)}$ exist? For the unit ball the result was obtained by Doob (1959) in [72]. The case of Lipschitz bounded domains was considered by Wu (1978) in [153].

**Theorem 3.2 (Relative Fatou's Theorem).** *If $f, g \geq 0$ are harmonic functions on a Lipschitz domain $D$, that is $\Delta f = 0$ on $D$, having the Martin representation $f(x) = \int_{\partial D} M_D(x, z)\mu(dz)$, $g(x) = \int_{\partial D} M_D(x, z)\nu(dz)$, where $M_D(x, z)$ is the Martin kernel of $D$ and $\mu, \nu$ are finite positive measures on $\partial D$. Then the nontangential (finite) limit*

$$\lim_{x \to Q \in \partial D} \frac{f(x)}{g(x)}$$

*exists a.s. with respect to the measure $\nu$.*

A natural question arises if both Fatou's theorems could be proved for $\alpha$-harmonic function, $0 < \alpha < 2$ (for the definition see the next section).

The following example exhibits that the behavior of the $\alpha$-harmonic functions, $\alpha < 2$, is less regular than in the classical case.

**Example 3.3.** *Let $\alpha < 2$. Suppose that $D$ is the open unit ball in $\mathbb{R}^d$. Then the function*

$$f(x) = (1 - |x|^2)^{\alpha/2 - 1}, \quad |x| < 1$$

$$f(x) = 0, \quad |x| \geq 1$$

*is $\alpha$-harmonic in $D$ and*

$$\lim_{x \to Q \in \partial D} f(x) = \infty.$$

Bass and You (2003) in [15] provide examples of positive bounded $\alpha$-harmonic functions on a half-space such that the nontangential limit does not exist at almost all points of the boundary. They also gave some sufficient conditions for the nontangential convergence.

There are probabilistic proofs of the Fatou theorem taking the advantage of the fact that for harmonic $f \geq 0$ and a Brownian motion $W_t$,

$$M_t = f(W_{t \wedge \tau_D})$$

is a positive supermartingale and $W_{t \wedge \tau_D}$ hits the boundary at the exiting time $\tau_D$.

Let us remark that for the $\alpha$-stable process, $\alpha < 2$, due to its jumping nature, this argument can not be applied. Nevertheless it turns out that the relative Fatou theorem for $\alpha$-harmonic functions holds at least for some points of the boundary but we need to impose some further assumptions on the functions (see Theorem 3.4). Namely we need to consider so called *singular* $\alpha$-harmonic functions, which are precisely defined in the next section.

## 3.2 Basic Definitions and Properties

We assume that $(X_t, P^x)$ is a rotation invariant (isotropic) $\alpha$-stable Lévy motion on $\mathbb{R}^d, d \geq 2$ with index $\alpha \in (0,2)$. That is it has independent stationary increments and its characteristic function is given by the following formula:

$$E^0 e^{iz \cdot X_t} = e^{-t|z|^\alpha}, \quad z \in \mathbb{R}^d.$$

For an open set $D \subset \mathbb{R}^d$ we define its first exit time $\tau_D = \inf\{t \geq 0 : X_t \notin D\}$. One of the basic objects in the potential theory of the stable process is the so-called $\alpha$-*harmonic measure* of $D$ given by

$$\omega_D^x(B) = P^x(X_{\tau_D} \in B), \ B \subset D^c, \ x \in \mathbb{R}^d.$$

Recall that an open $D$ is a Lipschitz set if there exist constants $R_0$ (*localization radius*) and $\lambda > 0$ (*Lipschitz constant*) such that for every $Q \in \partial D$ there is a function $F : \mathbb{R}^{d-1} \to \mathbb{R}$ and an orthonormal coordinate system $y = (y_1, ..., y_d)$ such that

$$D \cap B(Q, R_0) = \{y : y_d > F(y_1, ..., y_{d-1})\} \cap B(Q, R_0).$$

Moreover $F$ is Lipschitz with the Lipschitz constant not greater than $\lambda$.

If $F$ is differentiable and $\nabla F$ is Lipschitz with the Lipschitz constant not greater than $\lambda$ then $D$ is called a $C^{1,1}$ set.

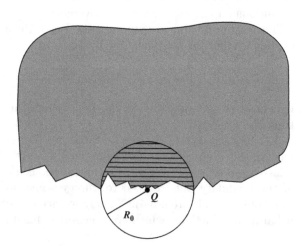

**Fig. 3.1** Typical
Lipschitz domain

If $D$ is a bounded Lipschitz set then $\omega_D^x(\partial D) = 0$ and $\omega_D^x$ has the density with respect to the Lebesgue measure called the *Poisson kernel* of $D$,

$$P_D(x,y), \ x \in D, y \in (\overline{D})^c.$$

When $D = B(0,r)$, $r > 0$, the Poisson kernel is given by an explicit formula:

$$P_r(x,y) = C_{d,\alpha} \left( \frac{r^2 - |x|^2}{|y|^2 - r^2} \right)^{\alpha/2} \frac{1}{|x-y|^d}, \quad |x| < r, |y| > r,$$

where $C_{d,\alpha} = \Gamma(d/2)\pi^{-d/2-1}\sin(\pi\alpha/2)$ (see [25]).

A Borel function $h$ on $\mathbb{R}^d$ is said to be $\alpha$-*harmonic* on $D$ if for each bounded open set $B$ with $\overline{B} \subset D$ and for $x \in B$ we have

$$h(x) = E^x h(X_{\tau_B}),$$

where the last integral is absolutely convergent. If $h \equiv 0$ on $D^c$ then it is called *singular* $\alpha$-*harmonic* on $D$. On the other hand $h$ is called *regular* $\alpha$-*harmonic* on $D$ if

$$h(x) = E^x h(X_{\tau_D}).$$

For example the harmonic measure $\omega_D^x(B)$ for a fixed $B$ is *regular* $\alpha$-*harmonic* on $D$ as a function of $x$.

We define the *Riesz potential* by the formula

$$U(x-y) = \int_0^\infty p_t(x-y)dt = \mathcal{A}_{d,\alpha}|x-y|^{\alpha-d},$$

where $p_t(x-y)$ is the transition density of the process $X_t$ and $\mathcal{A}_{d,\alpha}$ is a positive constant dependent on $d, \alpha$.

The *Green function* of $D$ is defined as

$$G_D(x,y) = \mathcal{A}_{d,-\alpha}(|x-y|^{\alpha-d} - E^x|y - X_{\tau_D}|^{\alpha-d}), \quad x,y \in D, x \neq y.$$

A very important result in the potential theory of the stable process is

**Boundary Harnack Principle (BHP).** *Let $D$ be a Lipschitz domain and $U$ an open subset of $\mathbb{R}^d$, $K$ a compact subset of $U$ such that $K \cap D \neq \emptyset$. There exists $c_o \in (0,\infty)$ such that for every $u, v \geqslant 0$, $\alpha$-harmonic in $D$ and vanishing continuously on $D^c \cap U$, with $u(x_0) = v(x_0)$ for fixed $x_0 \in K$ we have*

$$c_o v(y) \leqslant u(y) \leqslant c_o^{-1} v(y),$$

*for all $y \in K \cap D$.*

This result was proved by Bogdan [27]. See also [150].

Another important object of the potential theory is the so-called *Martin kernel*. It may be defined as the following limit, which exists due to **BHP**:

$$M(x,z) = \lim_{D \ni y \to z} \frac{G_D(x,y)}{G_D(x_0,y)}, \quad x \in D, z \in \partial D,$$

where $x_0 \in D$ is a reference point chosen at our convenience.

Every nonnegative singular $\alpha$-harmonic function on $D$ has a unique representation called the *Martin representation*. That is

$$f(x) = \int_{\partial D} M(x,z)\mu(dz),$$

where $\mu$ is a finite measure on $\partial D$ (see Chen and Song (1998) [50] or Bogdan (1999)[28]).

For the unit ball and $x_0 = 0$:

$$M(x,z) = \frac{(1 - |x|^2)^{\alpha/2}}{|x - z|^d}, \quad |x| < 1; |z| = 1.$$

Let us observe that taking the uniform measure on the unit sphere one can easily show that the function

$$f(x) = (1 - |x|^2)^{\alpha/2 - 1}, \quad |x| < 1; \quad f(x) = 0, \quad |x| \geq 1,$$

is an example of a singular $\alpha$-harmonic function on the unit ball. Recall that this is the function from Example 3.3.

## 3.3   Relative Fatou Theorem for $\alpha$-Harmonic Functions

Throughout the whole section we assume that $D$ is a Lipschitz bounded set unless stated otherwise. Suppose that $u(x), v(x)$ are two nonnegative singular $\alpha$-harmonic functions, $0 < \alpha < 2$, determined by two Borel finite measures concentrated on $\partial D$. That is

$$u(x) = \int_{\partial D} M(x,z)\mu(dz), \quad v(x) = \int_{\partial D} M(x,z)\nu(dz).$$

Next we can write the following decomposition of the measure $\mu(dz)$:

$$d\mu = f d\nu + d\mu_s,$$

where $\mu_s$ is singular to $\nu$ and $f \in L^1(\nu)$ is a positive function.

The first result about the boundary limit behavior of the ratio of $u(x)$ and $v(x)$ was obtained by Bogdan and Dyda (2003) in [34] for the special case $d\mu = f d\sigma$, where $\sigma$ is the Haussdorff surface measure concentrated on $\partial D$ for a bounded $C^{1,1}$ open set. They showed that for $\sigma$-almost every point $Q \in \partial D$ we have

$$\lim_{x \to Q} \frac{u(x)}{v(x)} = f(Q),$$

provided that the limit is taken nontangentially. Then that result was generalized independently to the general setting for Lipschitz sets by Michalik and Ryznar 2004 [124] and Kim (2006) [101]. As a matter of fact Kim in [101] studied even more general sets than Lipschitz: namely he considered so called $\kappa$-fat sets (see Remark 3.5 below)

**Theorem 3.4.** *For $\nu$-almost every point $Q \in \partial D$ we have*

$$\lim_{x \to Q} \frac{u(x)}{v(x)} = f(Q)$$

*as $x \to Q$ nontangentially.*

The methods used in [124] and [101] to prove Theorem 3.4 are different. While the proof in [124] is mostly analytical then [101] uses more probabilistic arguments. In the rest of this section we provide the outline of the proof from [124]. We start with reminding some estimates of the Martin kernel and then we provide the proof of the maximal estimate (see Lemma 3.6) which is the main tool in establishing Theorem 3.4.

### Useful Estimates for Martin Kernel

The basic estimates for $M$ were obtained by Jakubowski (2002) in [98]. Before stating this result we need to define points $A_{Q,r}$, $Q \in \partial D$, $r > 0$. Let $x_0 \in D$ be a fixed reference point. For $r \leq R_0/32$ we denote by $A_{Q,r}$ a point for which

$$B(A_{Q,r}, \kappa r) \subset B(Q,r) \cap D \qquad (3.1)$$

for a certain absolute constant $\kappa = \kappa(D) = 1/(2\sqrt{1+\lambda^2})$. For $r > R_0/32$ we set $A_{Q,r} = x_1$, where $x_1 \in D$ is such that $|x_0 - x_1| = R_0/4$. The picture below explains how to locate $A_{Q,r}$.

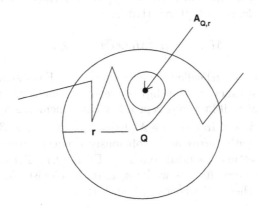

**Remark 3.5.** *There are non-Lipschitz sets for which the property (3.1) holds with some $\kappa$ and $0 < r \leq r_0$. Sets for which (3.1) holds are called $\kappa$-fat.*

The above mentioned result from [98] states that

$$c\frac{\phi(x)}{|x-z|^{d-\alpha}\phi^2(A_{z,|x-z|})} \leq M(x,z) \leq C\frac{\phi(x)}{|x-z|^{d-\alpha}\phi^2(A_{z,|x-z|})}, \qquad (3.2)$$

where $c, C$ depend on $d, \alpha, \lambda$ and

$$\phi(x) = \min(G_D(x,x_0), C_{d,\alpha}(R_0/4)^{\alpha-d}).$$

If $D$ is a $C^{1,1}$ domain then Kulczycki (1997) in [109], Chen and Song (1998) in [51] showed that

$$c\frac{\delta_x(D)^{\alpha/2}}{|x-z|^d} \leq M(x,z) \leq C\frac{\delta_x(D)^{\alpha/2}}{|x-z|^d},$$

where $\delta_x(D) = \text{dist}\,(\partial D, x)$.

To take advantage of (3.2) one has to estimate the function $\phi(x)$. This can be achieved by considering cones and their Martin kernels (see [5]).

An *unbounded circular cone* with vertex at $0 = (0,0,...,0)$ and symmetric with respect to the $d$-th axis is a set $\Gamma$ defined as

$$\Gamma = \{x : \eta \cdot |(x_1, x_2, ..., x_{d-1})| < x_d\},$$

where $\eta \in (-\infty, \infty)$, see Figure 3.2 below. The *aperture* of $\Gamma$ is the angle $\gamma = \arccos(\eta/\sqrt{1+\eta^2}) \in (0, \pi)$.

Let $\Gamma$ be a cone with vertex at $0$ and aperture $\gamma \in (0, \pi)$. Assume that $1 = (0,0,...,0,1)$. From Bogdan and Bañuelos (2003) [5] there exists a unique nonnegative function $M_\Gamma$ on $\mathbb{R}^d$ *the Martin kernel with pole at infinity* called such that $M_\Gamma(1) = 1$, $M_\Gamma \equiv 0$ on $\Gamma^c$ and $M_\Gamma$ is regular $\alpha$-harmonic on every open bounded subset of $\Gamma$. Moreover, $M_\Gamma$ is locally bounded on $\mathbb{R}^d$ and homogeneous of degree $\beta \in [0, \alpha)$, that is,

$$M_\Gamma(x) = |x|^\beta M_\Gamma(x/|x|), \quad x \in \Gamma.$$

Furthermore, $\beta = \beta(\Gamma, \alpha)$ called *the characteristic* of $\Gamma$ is a strictly decreasing function of $\gamma$. An exact relationship between the aperture and the characteristic of $\Gamma$ is not well known except a few cases including $\gamma = \pi/2$, that is when $\Gamma$ is a half-space. In this case $M_\Gamma(x) = x_d^{\alpha/2}, x \in \Gamma$, so $\beta = \alpha/2$.

For a cone $\Gamma$ with vertex at $Q$ (obviously isometric to some cone with vertex at $0$) we define a bounded cone $\Gamma_r = \Gamma \cap B(Q, r)$. For any $Q \in \partial D$, by the properties of Lipschitz sets we know that there exists $R_0 > 0$ such that for every $r \leq R_0$ there exist cones $\Gamma, \tilde{\Gamma}$ such that

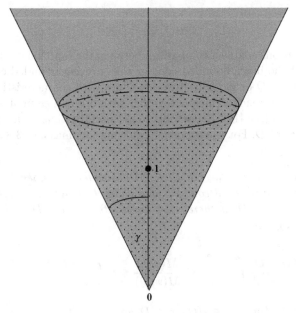

**Fig. 3.2** Bounded and unbounded circular cones

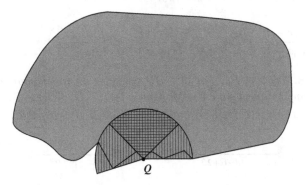

**Fig. 3.3** Inner cone and covering cone

$$\Gamma_r \subset D \cap B(Q, R_0) \subset \widetilde{\Gamma}_r.$$

Here $\beta, \widetilde{\beta}$ may depend locally on $\Gamma, r$. We will call $\Gamma_r$ an *inner bounded cone* and $\widetilde{\Gamma}_r$ a *covering bounded cone* (see Figure 3.3).

There are $r_0$ and $\beta_0, \widetilde{\beta}_0$ such that cones with characteristic $\beta_0 \geq \widetilde{\beta}_0$, respectively, are universal for every boundary point:

$$B(A_{Q,r_0}, \kappa r_0) \subset \Gamma_{r_0} \subset \widetilde{\Gamma}_{r_0}$$

and
$$\Gamma_{r_0} \subset B(Q, r_0) \cap D \subset \tilde{\Gamma}_{r_0}.$$

These observations enable a comparison between the function $\phi$ and the Martin kernels of the inner bounded cone or the covering bounded cone locally at the vicinity of $Q$ or globally if we use cones with characteristics $\beta_0 \geq \tilde{\beta}_0$. The Boundary Harnack Principle together with optimal estimates of Martin kernels for cones (see [123]) give tools to get useful estimates for the Martin kernel $M(x, z)$ of $D$. For the proofs of following Lemmas 3.6, 3.8 and 3.9 see [124].

**Lemma 3.6.** *Let $r \leq R_0$. For $Q \in \partial D$ consider an inner bounded cone $\Gamma_r$ and a covering bounded cone $\tilde{\Gamma}_r$ with characteristics $\beta, \tilde{\beta}$ . Moreover, let $|x - Q| \leq |x - z|$. Then there exist constants $c = c(r, D, \beta, \tilde{\beta})$ and $C = C(r, D, \beta, \tilde{\beta})$ such that*

$$c\left(\frac{|x - Q|}{|x - z|}\right)^{d-\alpha+2\beta} \leq \frac{M(x, z)}{M(x, Q)} \leq C\left(\frac{|x - Q|}{|x - z|}\right)^{d-\alpha+2\tilde{\beta}}.$$

**Corollary 3.7.** *Let $z, z' \in \partial D$, $x \in D$ and $|x - z'| \leq |x - z|$ . There are universal $\beta_0 \geq \tilde{\beta}_0$ both smaller than $\alpha$ such that there exist constants $c = c(R_0, D, \beta_0, \tilde{\beta}_0)$ and $C = C(R_0, D, \beta_0, \tilde{\beta}_0)$ such that*

$$c\left(\frac{|x - z'|}{|x - z|}\right)^{d-\alpha+2\beta_0} \leq \frac{M(x, z)}{M(x, z')} \leq C\left(\frac{|x - z'|}{|x - z|}\right)^{d-\alpha+2\tilde{\beta}_0}.$$

**Lemma 3.8.** *Let $r \leq R_0$. For $Q \in \partial D$ consider an inner bounded cone $\Gamma_r$ and a covering bounded cone $\tilde{\Gamma}_r$. Moreover, let $|z - Q| \leq 2|x - z|$. Then there exist constants $c = c(r, D, \beta, \tilde{\beta})$ and $C = C(r, D, \beta, \tilde{\beta})$ such that*

$$c\frac{\phi(x)}{|x - z|^{d-\alpha+2\tilde{\beta}}} \leq M(x, z) \leq C\frac{\phi(x)}{|x - z|^{d-\alpha+2\beta}}.$$

*Furthermore, if $\beta = \beta_0$ and $\tilde{\beta} = \tilde{\beta}_0$ then $c = c(R_0, D, \beta_0, \tilde{\beta}_0)$ and $C = C(R_0, D, \beta_0, \tilde{\beta}_0)$. In this case the above estimates hold for all $z \in \partial D, x \in D$.*

**Lemma 3.9.** *If $v(x) = \int_{\partial D} M(x, z)\nu(dz)$ then*

$$\liminf_{x \to Q} v(x) > 0$$

*for $\nu$ almost all $Q$ provided the limit is nontangential.*

As an immediate consequence of the above lemma and the well known fact

$$G_D(x, x_0) \to 0 \quad \text{if } x \to Q$$

we have that

$$\frac{G_D(x, x_0)}{v(x)} \to 0 \quad \text{as } x \to Q \tag{3.3}$$

for $\nu$ almost all $Q$ provided the limit is nontangential.

Now we can easily show that if $d\mu = f d\nu$ and $f$ is continuous at $Q$ then the ratio of $u$ and $v$ converges to $f$ for all such points $Q$ that (3.3) holds. To see this we define

$$I_1(x) = \int_{\partial D \cap \{|z-Q| \geq \varepsilon\}} |f(z) - f(Q)| M(x, z) \nu(dz),$$

$$I_2(x) = \int_{\partial D \cap \{|z-Q| < \varepsilon\}} |f(z) - f(Q)| M(x, z) \nu(dz).$$

and then

$$\left| \frac{u(x)}{v(x)} - f(Q) \right| \leq \frac{\int_{\partial D} |f(z) - f(Q)| M(x, z) \nu(dz)}{v(x)}$$

$$= \frac{I_1(x)}{v(x)} + \frac{I_2(x)}{v(x)}.$$

At first we estimate the term $\frac{I_1(x)}{v(x)}$. By Lemma 3.4,

$$M(x, z) \leq C \frac{\phi(x)}{|x - z|^{d-\alpha+2\beta_0}} \leq C \frac{G_D(x, x_0)}{|x - z|^{d-\alpha+2\beta_0}}.$$

If $|z - Q| \geq \varepsilon$ and $|x - Q| \leq \varepsilon/2$ then $|x - z| \geq \varepsilon/2$. This implies

$$M(x, z) \leq C(\varepsilon) G_D(x, x_0).$$

Hence

$$I_1(x) = \int_{\partial D \cap \{|z-Q| \geq \varepsilon\}} |f(z) - f(Q)| M(x, z) \nu(dz)$$

$$\leq C(\varepsilon) |\mu| G_D(x, x_0) \int_{\partial D} |f(z) - f(Q)| \nu(dz),$$

which together with (3.3) shows that

$$\frac{I_1(x)}{v(x)} \to 0, \quad \text{as } x \to Q. \tag{3.4}$$

Observe that the above limit holds without the continuity assumption.

Next, note that

$$I_2(x) \leq \sup_{|z-Q|\leq\varepsilon} |f(z) - f(Q)|v(x)$$

hence

$$\frac{I_2(x)}{v(x)} \leq \sup_{|z-Q|\leq\varepsilon} |f(z) - f(Q)| \to 0, \ as \ \varepsilon \to 0$$

by the continuity of $f$ at $Q$.

To provide the proof of the relative Fatou theorem in the general case we need the following crucial lemma.

**Lemma 3.10 (Nontangential Maximal Estimate).** *Suppose that*

$$u(x) = \int_{\partial D} M(x, z)\mu(dz), \quad v(x) = \int_{\partial D} M(x, z)\nu(dz).$$

*For any $x \in D$, $Q \in \partial D$ and $t > 0$ such that $|x - Q| \leq t\delta_x(D)$ there exist constants $C = C(t, Q)$, $c = c(t, Q)$ such that*

$$\frac{u(x)}{v(x)} \leq C \sup_{r>0} \frac{\mu(B(Q, r))}{\nu(B(Q, r))}.$$

*Proof.* For $n \geq 1$ set

$$B_n = B(Q, 2^n|x - Q|)$$

and

$$A_1 = B_1, \quad A_n = B_n \setminus B_{n-1}, n \geq 2.$$

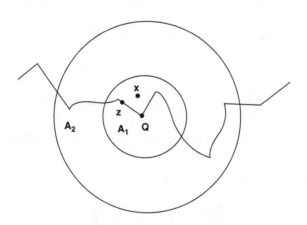

If $z \in A_1$ then

$$|x - Q|/t \leq \delta_x \leq |x - z| \leq 3|x - Q|.$$

If $z \in A_n$ then

$$(2^{n-1} - 1)|x - Q| \leq |x - z| \leq (2^n + 1)|x - Q|.$$

Define $a_n = \sup_{z \in A_n} M(x, z)$, $z \in A_n$ and observe that by Lemma 3.2

$$a_n \approx M(x, z), z \in A_n.$$

If $z' \in A_n$ and $z \in A_k$, $k > n$ then

$$|x - z'| \leq 4|x - z|.$$

Hence by Lemma 3.2 we obtain

$$M(x, z) \leq CM(x, z')$$

for some $C \geq 1$ dependent on $t$ and $Q$. It implies that $a_k \leq Ca_n$ for $k > n$ and for

$$b_n = \sup_{k \geq n} a_k, \ n \geq 1,$$

this yields

$$a_n \leq b_n \leq Ca_n.$$

Then we obtain

$$u(x) = \int_{\partial D} M(x, z)\mu(dz) = \sum_{n=1}^{n_0} \int_{A_n} M(x, z)\mu(dz)$$

$$\approx \sum_{n=1}^{n_0} a_n\mu(A_n)$$

$$\approx \sum_{n=1}^{n_0} b_n\mu(A_n),$$

where $n_0$ be the smallest index for which $2^{n_0}|x - Q| \geq \text{diam}(D)$. Denote

$$R = \sup_{r>0} \frac{\mu(B(Q, r))}{\nu(B(Q, r))}$$

Now summation by parts yields (note that $b_n$ is nonincreasing):

$$u(x) \approx \sum_{n=1}^{n_0} b_n\mu(A_n) =$$

$$= \sum_{n=2}^{n_0} (b_{n-1} - b_n)\frac{\mu(B_n)}{\nu(B_n)}\nu(B_n) + b_{n_0}\frac{\mu(B_{n_0})}{\nu(B_{n_0})}\nu(B_{n_0})$$

$$\leq R \left( \sum_{n=2}^{n_0} (b_{n-1} - b_n)\nu(B_n) + b_{n_0}\nu(B_{n_0}) \right)$$

$$= R \sum_{n=1}^{n_0} b_n \nu(A_n)$$

$$\approx Rv(x). \qquad \qquad \square$$

Now we can complete the proof of Theorem 3.4. Let $\tilde{\mu}(dz) = |f(z) - f(Q)|\nu(dz) + \mu_s(dz)$. Then

$$\left| \frac{u(x)}{v(x)} - f(Q) \right| \leq \int_{\partial D} M(x,z)\tilde{\mu}(dz)/v(x)$$

$$\leq \int_{\partial D \cap \{|z-Q| \geq \varepsilon\}} M(x,z)\tilde{\mu}(dz)/v(x)$$

$$+ \int_{\partial D \cap \{|z-Q| < \varepsilon\}} M(x,z)\tilde{\mu}(dz)/v(x).$$

The first term converges to 0, by the same argument as leading to (3.4), if $x \to Q$ and (3.3) holds. To take care of the second term we apply the Maximal Inequality to arrive at

$$\int_{\partial D \cap \{|z-Q| < \varepsilon\}} M(x,z)\tilde{\mu}(dz)/v(x)$$

$$\leq C \left( \sup_{r \leq \varepsilon} \frac{\int_{\partial D \cap B(Q,r)} |f(z) - f(Q)|\nu(dz)}{\nu(B(Q,r))} + \sup_{r \leq \varepsilon} \frac{\mu_s(B(Q,r))}{\nu(B(Q,r))} \right)$$

Next, observe that the set of points $Q$ for which the limit

$$\lim_{\varepsilon \to 0} \left( \sup_{r \leq \varepsilon} \frac{\int_{\partial D \cap B(Q,r)} |f(z) - f(Q)|\nu(dz)}{\nu(B(Q,r))} + \sup_{r \leq \varepsilon} \frac{\mu_s(B(Q,r))}{\nu(B(Q,r))} \right) \neq 0 \quad (3.5)$$

is of $\nu$ measure zero. This completes the sketch of the proof of Theorem 3.4.

Note that the advantage of the above proof is that we can identify the points $Q \in \partial D$ for which the ratio converges. Namely these are points for which (3.3) and (3.5) hold.

## 3.4 Extensions to Other Processes

At first we consider the relativistic $\alpha$-stable process $0 < \alpha < 2$. This is a Lévy process which characteristic function is of the form

$$E^0 e^{iz \cdot Y_t} = e^{-t((|z|^2 + m^{2/\alpha})^{\alpha/2} - m)}, \quad z \in \mathbb{R}^d,$$

where $m > 0$ is a parameter (see also Example 5.9 below). For $m = 0$ it reduces to the isotropic $\alpha$-stable process. The generator of this process is given by

$$A_m = m - \left(-\Delta + m^{2/\alpha}\right)^{\alpha/2}.$$

To explain the name *relativistic* observe that if $\alpha = 1$ the generator is related to the kinetic energy of the relativistic particle of mass $m$, see [45]. If $\nu_Y$ is the Lévy measure of $Y_t$ and $\nu_X$ is the Lévy measure of $X_t$ (isotropic stable) then (see [134]) $\nu_X - \nu_Y$ is a finite positive measure, which is equivalent to the fact that $A_m - A_0$ is a bounded operator. It suggests that the potential theory of such a process should bear a lot of similarity with the potential theory of the stable process. This is the case if we study properties of the process killed after exiting a bounded open set.

**Theorem 3.11 (Green Function Estimates).** *Let $D$ be an open Lipschitz bounded set. There is a positive constant $C = C(D)$ such that*

$$C^{-1}G_D^X(x,y) \leq G_D^Y(x,y) \leq CG_D^X(x,y), \quad x,y \in D, \tag{3.6}$$

*where $G_D^X$, $G_D^Y$ are corresponding Green functions of $D$.*

**Remark 3.12.** *The comparison above may not hold for unbounded sets. In the paper [83] optimal estimates for the Green function of a half-space are derived and they exhibit that the Green function is comparable to the Brownian Green function for the points away from the boundary and not too close to each other.*

For a bounded open $C^{1,1}$ set $D$ the comparison (3.6) was proved by Ryznar (2002) [134] and by another method by Chen and Song (2003) [53]. Later on that result was proved for bounded open Lipschitz sets by Grzywny, Ryznar (2006) [82] and for $\kappa$-fat sets by Kim and Lee (2006) [102].

This estimate plus some other properties of the Lévy measure of $Y$ will allow to study potential theory for the relativistic process very much the same as for the stable process provided that $D$ is bounded and Lipschitz. For example one can prove the existence of the Martin kernel of $D$ for the process $Y$ and show that relative Fatou's theorem holds for nonnegative functions defined by the Martin kernel $M_D^Y(x,z)$:

$$h(x) = \int_{\partial D} M_D^Y(x,z)\mu(dz).$$

Such a result for example was proved in [101] and [102], where a class of nonlocal Feynman-Kac transforms of the stable process was studied. The relativistic stable process is included in this class. It is easy to see that the proof presented in the previous section will go through in the relativistic case since due to Theorem 3.11 we are able to derive all needed estimates for the Martin kernel of $Y$.

Theorem 4.1 suggests that we should expect a similar result if we take as $Y_t$ any Lévy process with its Lévy measure sufficiently close to the Lévy measure $\nu_X$. Let

$$E^0 e^{iz \cdot Y_t} = e^{-t \int_{\mathbb{R}^d} (1 - \cos(z \cdot w)) \nu_Y(dw)},$$

where $\nu_Y$ is the Lévy measure of $Y_t$ and we assume that

$$\nu_Y - \nu_X \quad \text{is a finite signed measure.}$$

**Question.** Is there a positive constant $C$ such that

$$C^{-1} G_D^X(x, y) \le G_D^Y(x, y) \le C G_D^X(x, y), \quad x, y \in D?$$

**Theorem 3.13.** *Suppose that $D$ is an open bounded set. If $\nu_X \ge \nu_Y$ then*

$$G_D^Y(x, y) \le C G_D^X(x, y).$$

*If $\nu_X \le \nu_Y$ then*

$$G_D^X(x, y) \le C G_D^Y(x, y)$$

This result was proved by Grzywny and Ryznar (2006) in [82]. Under the additional assumption about the densities $\nu_X(x), \nu_Y(x)$ of the corresponding Lévy measures:

$$|\nu_X(x) - \nu_Y(x)| \le C |x|^{-d+\varrho}, \quad |x| \le 1, \quad \varrho > 0,$$

the answer for the raised question is positive but we need to assume that $D$ is bounded Lipschitz and connected (see [82]). Namely there is a constant $C$ such that

$$C^{-1} G_D^X(x, y) \le G_D^Y(x, y) \le C G_D^X(x, y), \quad x, y \in D.$$

The assumption that $D$ is connected is not to avoid, since the Lévy measure $\nu_Y$ can be concentrated on some neighborhood of the origin so the process can only make small jumps, so $G_D(x, y)$ might be 0 if $D$ has components far away from each other. This will not happen for the stable case.

For open bounded $\kappa$-fat sets the result can be deduced under some additional assumptions on the behavior of the Lévy measure at $\infty$ from the results of Kim and Lee (2006) [102].

The comparability of the Green functions allows to prove many properties similar to those possessed by the stable process. Again we can expect the relative Fatou theorem holds however some further assumptions are necessary. For example for the truncated stable process, that is if $\nu_Y = \nu_X|_{B(0,R)}$, $R > 0$, the relative Fatou theorem is true for connected bounded $\kappa$-fat sets. This is a recent result obtained by Kim and Song in [104](see also [103]).

# Chapter 4
# Eigenvalues and Eigenfunctions for Stable Processes

by T. Kulczycki

## 4.1 Introduction

Let $X_t$ be a symmetric $\alpha$-stable process in $\mathbb{R}^d$, $\alpha \in (0, 2]$. This is a process with independent and stationary increments and characteristic function $E^0 e^{i\xi X_t} = e^{-t|\xi|^\alpha}$, $\xi \in \mathbb{R}^d$, $t > 0$. We will use $E^x$, $P^x$ to denote the expectation and probability of this process starting at $x$, respectively. By $p(t, x, y) = p_t(x - y)$ we will denote the transition density of this process. That is,

$$P^x(X_t \in B) = \int_B p(t, x, y) \, dy.$$

When $\alpha = 2$ the process $X_t$ is just the Brownian motion in $\mathbb{R}^d$ running at twice the speed. That is, if $\alpha = 2$ then

$$p^{(2)}(t, x, y) = \frac{1}{(4\pi t)^{d/2}} e^{\frac{-|x-y|^2}{4t}}, \quad t > 0, \; x, y \in \mathbb{R}^d. \tag{4.1}$$

When $\alpha = 1$, the process $X_t$ is the Cauchy process in $\mathbb{R}^d$ whose transition densities are given by

$$p^{(1)}(t, x, y) = \frac{c_d \, t}{(t^2 + |x - y|^2)^{(d+1)/2}}, \quad t > 0, \; x, y \in \mathbb{R}^d, \tag{4.2}$$

where

$$c_d = \Gamma((d + 1)/2) / \pi^{(d+1)/2}.$$

We are mainly interested in the eigenvalues and eigenfunctions of the semigroup $\{P_t^D\}_{t \geq 0}$ of the process $X_t$ killed upon exiting a domain $D \subset \mathbb{R}^d$.

Let $D \subset \mathbb{R}^d$ be a domain with finite Lebesgue measure $m(D) < \infty$ and

$$\tau_D = \inf\{t \geq 0 : X_t \notin D\}$$

K. Bogdan et al., *Potential Analysis of Stable Processes and its Extensions*, Lecture Notes in Mathematics 1980, DOI 10.1007/978-3-642-02141-1_4, © Springer-Verlag Berlin Heidelberg 2009

be the first exit time of $D$. By $\{P_t^D\}_{t \geq 0}$ we denote the semigroup on $L^2(D)$ of $X_t$ killed upon exiting $D$. That is,

$$P_t^D f(x) = E^x(f(X_t), \tau_D > t), \quad x \in D, \ t > 0, \ f \in L^2(D).$$

The semigroup has transition densities $p_D(t, x, y)$ satisfying

$$P_t^D f(x) = \int_D p_D(t, x, y) f(y) \, dy.$$

The kernel $p_D(t, x, y)$ is strictly positive symmetric and

$$p_D(t, x, y) \leq p(t, x, y) \leq c_{\alpha, d} \, t^{-d/\alpha}, \quad x, y \in D, \ t > 0.$$

It follows that

$$\int_D \int_D p_D^2(t, x, y) \, dx \, dy \leq c_{\alpha, d}^2 \, t^{-2d/\alpha} m(D)^2 < \infty$$

so for any $t > 0$ the operator $P_t^D$ is a Hilbert-Schmidt operator. Hence for any $t > 0$ the operator $P_t^D$ is compact. From the general theory of semigroups it follows that there exists an orthonormal basis of eigenfunctions $\{\varphi_n\}_{n=1}^{\infty}$ for $L^2(D)$ and corresponding eigenvalues $\{\lambda_n\}_{n=1}^{\infty}$ satisfying

$$0 < \lambda_1 < \lambda_2 \leqslant \lambda_3 \leqslant \dots$$

with $\lambda_n \to \infty$ as $n \to \infty$. That is, the pair $\{\varphi_n, \lambda_n\}$ satisfies

$$P_t^D \varphi_n(x) = e^{-\lambda_n t} \varphi_n(x), \quad x \in D, \ t > 0. \tag{4.3}$$

We also have

$$p_D(t, x, y) = \sum_{n=1}^{\infty} e^{-\lambda_n t} \varphi_n(x) \varphi_n(y).$$

The eigenfunctions $\varphi_n$ are continuous and bounded on $D$. In addition, $\lambda_1$ is simple and the corresponding eigenfunction $\varphi_1$, often called the ground state eigenfunction, is strictly positive on $D$. For more general properties of the semigroups $\{P_t^D\}_{t \geq 0}$, see [79], [24], [54].

Let $\mathcal{A}$ be the infinitesimal generator of the semigroup $\{P_t^D\}_{t \geq 0}$. We have

$$\mathcal{A}\varphi_n(x) = \lim_{t \to 0} \frac{P_t^D \varphi_n(x) - \varphi_n(x)}{t}$$

$$= \lim_{t \to 0} \frac{e^{-\lambda_n t} - 1}{t} \varphi_n(x)$$

$$= -\lambda_n \varphi_n(x).$$

For any $t > 0$ we also have

$$|\varphi_n(x)| = e^{\lambda_n t} \left| \int_D p_D(t, x, y) \varphi_n(y) \, dy \right|$$
$$\leq e^{\lambda_n t} ||\varphi_n||_\infty \int_D p_D(t, x, y) \, dy$$
$$= e^{\lambda_n t} ||\varphi_n||_\infty P^x(\tau_D > t) \to 0,$$

when $x \to x_0 \in \partial D$ and $x_0$ is a *regular* point. We say that $x_0 \in \partial D$ is a regular point when $P_0^x(\tau_D = 0) = 1$. It may be shown that if $D$ satisfies the outer cone condition at $x_0 \in \partial D$ then $x_0$ is regular.

It follows that if all points on $\partial D$ are regular we have

$$\begin{cases} \mathcal{A}\varphi_n(x) = -\lambda_n(x), \ x \in D \\ \varphi_n(x) = 0, \qquad x \in \partial D. \end{cases}$$

When $\alpha = 2$ and $X_t$ is Brownian motion the semigroup $\{P_t^D\}_{t \geq 0}$ is just the *heat semigroup*. Eigenvalues and eigenfunctions are the solutions of the *eigenvalue problem for the Dirichlet Laplacian* :

$$\begin{cases} \Delta\varphi_n(x) = -\lambda_n(x), \ x \in D \\ \varphi_n(x) = 0, \qquad x \in \partial D. \end{cases}$$

For example when $D = (0, \pi)$

$$\varphi_n(x) = \sqrt{2/\pi} \sin(nx), \quad \lambda_n = n^2.$$

Indeed
$$\sin''(nx) = -n^2 \sin(nx), \quad \sin(0) = \sin(n\pi) = 0.$$

In particular the first eigenfunction is strictly positive on $D = (0, \pi)$

$$\varphi_1(x) = \sqrt{2/\pi} \sin(x).$$

The second eigenfunction changes the sign once on $D = (0, \pi)$

$$\varphi_2(x) = \sqrt{2/\pi} \sin(2x).$$

Let us observe that eigenfunctions in this example satisfy the following property: $n$th eigenfunction $\varphi_n$ has exactly $n$ *nodal domains*. A nodal domain for an eigenfunction $\varphi$ is any connected component of a set on which $\varphi$ has a constant sign.

In general, the eigenfunctions of the eigenvalue problem for the Dirichlet Laplacian satisfies the Courant-Hilbert nodal domain theorem, which states that the $n$-th eigenfunction has no more than $n$ nodal domains.

When $\alpha \in (0, 2)$ eigenfunctions and eigenvalues satisfy

$$\begin{cases} -(-\Delta)^{\alpha/2}\varphi_n(x) = -\lambda_n(x), \ x \in D \\ \varphi_n(x) = 0, \qquad\qquad\quad x \in D^c, \end{cases} \qquad (4.4)$$

where

$$-(-\Delta)^{\alpha/2}f(x) = \lim_{\varepsilon\downarrow 0} \mathcal{A}_{d,-\alpha} \int_{|x-y|>\varepsilon} \frac{f(y) - f(x)}{|y - x|^{d+\alpha}}\,dy.$$

with $\mathcal{A}_{d,\gamma} = \Gamma((d - \gamma)/2)/(2^\gamma \pi^{d/2}|\Gamma(\gamma/2)|)$. Nevertheless, in the case $\alpha \in (0, 2)$ the operator $-(-\Delta)^{\alpha/2}$ is a pseudodifferential nonlocal operator and it is very difficult to obtain some properties of the semigroup and its eigenfunctions and eigenvalues using equations (4.4).

## 4.2   Intrinsic Ultracontractivity (IU)

Intrinsic ultracontractivity is the property of the semigroup which is very useful in studying spectral properties of the eigenvalues and eigenfunctions cf. [6].

**Definition 4.1.** *The semigroup $\{P_t^D\}_{t\geq 0}$ is intrinsically ultracontractive (IU) when for any $t > 0$ there exists $c_t$ such that for any $x, y \in D$ we have*

$$\forall t > 0 \ \exists c_t < \infty \ \forall x, y \in D \quad p_D(t, x, y) \leq c_t\varphi_1(x)\varphi_1(y).$$

There are many other well known equivalent definitions of **IU**. We present probabilistic definition of **IU** which will be very useful in studying semigroups for stable processes.

**Proposition 4.1.** *The semigroup $\{P_t^D\}_{t\geq 0}$ is IU if and only if*

$$\exists K \subset\subset D \ \forall t > 0 \ \exists c = c_{t,K,D} \ \forall x \in D \quad P^x(\tau_D > t) \leq cP^x(\tau_D > t, X_t \in K).$$

The following very important result is proved in [110].

**Theorem 4.2.** *When $\alpha \in (0, 2)$ and $D \subset \mathbb{R}^d$ is bounded then $\{P_t^D\}_{t\geq 0}$ is IU.*

*Proof.* (idea) Let us fix $K = B(x_0, r)$ such that $\overline{K} \subset D$ and let us denote $K_1 = B(x_0, r/2)$. We will use $c$ to denote positive constant which depend on $t > 0$, $d$, $D$, $K$ and $\alpha$. This constant may change its value from line to line. We have for $x \in D$

$$P^x(\tau_D > t, X_t \in K)$$
$$\geq P^x(\tau_{D\setminus K_1} < t, X(\tau_{D\setminus K_1}) \in K_1, X_s \text{ in } K \text{ for all } s \in (\tau_{D\setminus K_1}, \tau_{D\setminus K_1} + t))$$
$$\geq cP^x(\tau_{D\setminus K_1} < t, X(\tau_{D\setminus K_1}) \in K_1).$$

By the Ikeda-Watanabe formula ([96]) it equals for $x \in D \setminus K_1$

$$c \int_D \int_0^t p_{D\setminus K_1}(s, x, y) \, ds \frac{\mathcal{A}_{d,-\alpha}}{|y - z|^{d+\alpha}} \, dy \, dz.$$

We have $|y - z|^{-d-\alpha} \geq (\text{diam}(D))^{-d-\alpha}$ so the above expression is bounded below by

$$c \int_D \int_0^t p_{D\setminus K_1}(s, x, y) \, dy \, ds$$
$$\geq c \int_0^t P^x(\tau_{D\setminus K_1} > s) \, ds$$
$$\geq ctP^x(\tau_{D\setminus K_1} > t).$$

It follows that for $x \in D$

$$P^x(\tau_D > t) \leq P^x(\tau_{D\setminus K_1} \geq t) + P^x(\tau_{D\setminus K_1} < t, X(\tau_{D\setminus K_1}) \in K_1)$$
$$\leq cP^x(\tau_D > t, X_t \in K). \qquad \square$$

We have
$$p_D(t, x, y) = \int_0^\infty e^{-\lambda_n t} \varphi_n(x) \varphi_n(y).$$

It follows that

$$\frac{p_D(t, x, y)}{e^{-\lambda_1 t} \varphi_1(x) \varphi_1(y)} = 1 + \sum_{n=2}^\infty e^{-(\lambda_n - \lambda_1)t} \frac{\varphi_n(x) \varphi_n(y)}{\varphi_1(x) \varphi_1(y)}.$$

**Proposition 4.3 (Consequences of IU).** *Let $\alpha \in (0, 2)$ and $D \subset \mathbb{R}^d$ is a bounded open set. Then*
   *i)*
$$\lim_{t \to \infty} \frac{p_D(t, x, y)}{e^{-\lambda_1 t} \varphi_1(x) \varphi_1(y)} = 1.$$

*ii) $\exists C = C(D, \alpha) \ \forall t \geq 1$*

$$e^{-(\lambda_2 - \lambda_1)t} \leq \sup_{x,y \in D} \left| \frac{p_D(t, x, y)}{e^{-\lambda_1 t} \varphi_1(x) \varphi_1(y)} \right| \leq Ce^{-(\lambda_2 - \lambda_1)t}$$

The next theorem proved in [110] gives asymptotic of $\varphi_1$.

**Theorem 4.4.** *Let $\alpha \in (0,2)$ and $D \subset \mathbb{R}^d$ is a bounded open set. Then there exist constants $C_1 = C_1(D,\alpha)$, $C_2 = C_2(D,\alpha)$ such that for all $x \in D$*

$$C_1 E^x(\tau_D) \leq \varphi_1(x) \leq C_2 E^x(\tau_D).$$

This theorem implies that for $\alpha \in (0,2)$ when $D = B(0,r)$ is a ball in $\mathbb{R}^d$ we have

$$\varphi_1(x) \approx E^x(\tau_D) = c_{\alpha,d}(r^2 - |x|^2)^{\alpha/2},$$

when $D = B(0,r)$ is a ball in $\mathbb{R}^d$,

$$\varphi_1(x) \approx E^x(\tau_D) \approx \delta_D^{\alpha/2}(x),$$

when $D$ is a bounded $C^{1,1}$ domain $(\delta_D(x) = \mathrm{dist}(x,\partial D))$,

$$c_1 \delta_D^{\gamma_1}(x) \leq \varphi_1(x) \leq c_2 \delta_D^{\gamma_2}(x),$$

when $D$ is a bounded Lipschitz domain, $c_1, c_2, \gamma_1, \gamma_2 > 0$ depend on $D$ and $\alpha$.

Intrinsic ultracontractivity may be investigated for some other semigroups e.g. Feynman-Kac semigroups in $\mathbb{R}^d$ with Schrödinger generators $H_0 - V$ where $V : \mathbb{R}^d \to \mathbb{R}$ is a potential. Recently the following result has been obtained for relativistic $\alpha$-stable processes. The relativistic $\alpha$-stable process in $\mathbb{R}^d$ is a Markov process with independent and homogeneous increments and the characteristic function of the form

$$E^0(\exp(i\xi X_t)) = \exp\left(-t((m^{2/\alpha} + |\xi|^2)^{\alpha/2} - m)\right),$$

$m > 0$.

The following result was proved in [112].

**Theorem 4.5.** *Let $\alpha \in (0,2)$, $d > \alpha$. Let $X_t$ be a relativistic $\alpha$-stable process and $\{T_t\}_{t \geq 0}$ a Feynman-Kac semigroup for $X_t$ with generator $-((-\Delta + m^{2/\alpha})^{\alpha/2-m}) - V$, where potential $V(x) = |x|^\beta$, $\beta > 0$. We have*

*i) $T_t$ are compact for any $\beta > 0$,*
*ii) $\{T_t\}_{t \geq 0}$ is* **IU** *if and only if $\beta > 1$,*
*iii) for $\beta > 1$*

$$\varphi_1(x) \approx \frac{\exp(-m^{1/\alpha}|x|)}{(|x|+1)^{(d+\alpha+2\beta+1)/2}}.$$

## 4.3 Steklov Problem

**IU** gives some general properties of the first and consecutive eigenfunctions for the semigroup of symmetric $\alpha$-stable processes. Nevertheless we do not know many fine properties of eigenvalues and eigenfunctions which are known for eigenvalues and eigenfunctions of Dirichlet Laplacian ($\alpha \in (0,2)$). Even in the most simple geometric case when $D = (-1,1)$ we do not know almost any properties.

A very simple question is the following. What can be said about the second eigenfunction $\varphi_2$? How many zeroes has this function on $(-1,1)$?

Some answers were obtained for the Cauchy process (symmetric $\alpha$-stable process for $\alpha = 1$). This was done in [7] by using the connection of the spectral problem for this process and the *Steklov problem*.

The connection between the eigenvalue problem for the Cauchy process and the Steklov problem arises as follows.

Let us consider the following construction. Let $D = (-1,1)$ and $\varphi_n, \lambda_n$ be eigenvalues and eigenfunctions for the semigroup of the Cauchy process (i.e. $\alpha$-stable process for $\alpha = 1$) killed on exiting $D$. $\varphi_n$ are defined on $D$ but let us defined them on $\mathbb{R}$ by putting $\varphi_n(x) = 0$ for $x \in D^c$. Since $|\varphi_n(x)| \le c_n \varphi_1(x) \approx (1 - |x|^2)^{1/2}$ on $D$ we get that $\varphi_n$ is continuous on $\mathbb{R}$.

Let
$$H = \{(x,t) : x \in \mathbb{R}^d, t \ge 0\}$$

be the upper half-space. Let us also denote $H_+ = \{(x,t) : x \in \mathbb{R}^d, t > 0\}$. We define functions $u_n(x,t)$ for $(x,t) \in H$ in the following way. For $t = 0$ we put $u_n(x,0) = \varphi_n(x)$. For $t > 0$ we put

$$u_n(x,t) = \int_{\mathbb{R}^d} p(t,x,y)\varphi_n(y)\, dy,$$

where $p(t,x,y)$ is the transition density for the Cauchy process ($\alpha$-stable process for $\alpha = 1$). Let us recall that

$$p(t,x,y) = \frac{1}{\pi} \frac{t}{t^2 + (x-y)^2}, \quad t > 0, \quad x,y \in \mathbb{R}.$$

Since $\varphi_n$ are continuous on $\mathbb{R}$ we obtain that $u_n$ are continuous on $H$.

A very important fact is that the transition density for the Cauchy process $p(t,x,y)$ is the Poisson kernel for upper half-space $H$ for classical harmonic measure. More precisely $p(t,x,y)$ is the Poisson kernel for points $(y,t) \in H_+$, $(x,0) \in \partial H$. It follows that $u_n(x,t)$ is harmonic for $(x,t) \in H_+$ for Laplacian $\Delta = \frac{\partial^2}{\partial x^2} + \frac{\partial^2}{\partial t^2}$ in $\mathbb{R}^2$.

It can also be shown that

$$\frac{\partial u_n}{\partial t}(x,0) = -\lambda_n u_n(x,0).$$

Heuristically it can be verified as follows

$$\frac{\partial u_n}{\partial t}(x,0) = \lim_{t \to 0^+} \frac{u_n(x,t) - u_n(x,0)}{t}$$
$$= \lim_{t \to 0^+} \int_{\mathbb{R}^d} \frac{p(t,x,y)(\varphi_n(y) - \varphi_n(x))}{t}\, dy$$

$$= \lim_{t \to 0^+} c_d \int_{\mathbb{R}^d} \frac{\varphi_n(y) - \varphi_n(x)}{(t^2 + |x - y|^2)^{\frac{d+1}{2}}} \, dy$$

$$= -(-\Delta)^{1/2} \varphi_n(x)$$

$$= -\lambda_n \varphi_n(x).$$

The precise proof may be find in [7] (proof of Theorem 1.1).

Therefore $u_n(x,t)$ satisfy

$$\Delta u_n(x,t) = 0; \qquad\qquad (x,t) \in H_+, \qquad\qquad (4.5)$$

$$\frac{\partial u_n}{\partial t}(x,0) = -\lambda_n u_n(x,0); \qquad x \in D \qquad\qquad (4.6)$$

$$u_n(x,0) = 0; \qquad\qquad x \in D^c. \qquad\qquad (4.7)$$

This problem is known as *the mixed Steklov problem*.

Using this connection the following result has been obtained [7].

**Theorem 4.6.** *Let* $\alpha = 1$, $D = (-1,1)$ *and let* $\varphi_n$, $\lambda_n$ *be eigenvalues and eigenfunctions for the semigroup of the Cauchy process (i.e.* $\alpha$-stable process *for* $\alpha = 1$) *killed on exiting* $D$. *Then we have*

*i)* $1 \leq \lambda_1 \leq 1.17$, $\varphi_1$ *is positive, symmetric, increasing on* $(-1,0)$, *decreasing on* $(0,1)$ *and concave on* $(-1,1)$,

*ii)* $2 \leq \lambda_2 \leq \pi$, $\varphi_2$ *is antisymmetric and (up to a sign) negative and convex on* $(-1,0)$, *positive and concave on* $(0,1)$,

*iii)* $3.4 \leq \lambda_3 \leq 3\pi/2$, $\varphi_3$ *is symmetric and has 2 zeroes on* $(-1,1)$,

*iv) the spectral gap satisfies* $\lambda_2 - \lambda_1 \geq \lambda_1 \geq 1$,

*v)* $\varphi_n$ *has no more than* $2n - 2$ *zeroes on* $(-1,1)$.

The Steklov problem is known for many years and belongs to classical problems in spectral theory, although is less known then Dirichlet or Neumann problems. In the famous R. Courant and D. Hilbert book [63], it appears as the spectral problem in boundary conditions.

If $\Omega$ is a bounded domain in $\mathbb{R}^d$, we write its boundary $\partial\Omega$ as the disjoint union of two pieces, $(\partial\Omega)_1$ and $(\partial\Omega)_2$, and the classical *"mixed Steklov" eigenvalue problem* ([88], [68], [70]) is the following mixed boundary value problem:

$$\Delta u_n(z) = 0; \qquad\qquad z \in \Omega, \qquad\qquad (4.8)$$

$$\frac{\partial u_n}{\partial \nu}(z) = -e_n u_n(z); \qquad z \in (\partial\Omega)_1. \qquad\qquad (4.9)$$

$$u_n(z) = 0; \qquad\qquad z \in (\partial\Omega)_2, \qquad\qquad (4.10)$$

where $\Delta = \sum_{i=1}^d \frac{\partial^2}{\partial x_i^2}$ and $\frac{\partial}{\partial \nu}$ is the inner normal derivative. The basic difference between our Steklov problems and the classical one in that our domain is unbounded.

The problem similar to the Steklov problem is considered also in hydro-dynamics. Let us consider the following problem.

$$\Delta u_n(x) = 0; \qquad\qquad x \in \Omega, \qquad\qquad (4.11)$$

$$\frac{\partial u_n}{\partial \nu}(x) = -\lambda_n u_n(x); \qquad x \in F. \qquad\qquad (4.12)$$

$$\frac{\partial u_n}{\partial \nu}(x) = 0; \qquad\qquad x \in B, \qquad\qquad (4.13)$$

where $\Omega \subset \mathbb{R}^3_- = \{x = (x_1, x_2, x_3) : x_3 < 0\}$, $F = \partial\Omega \cap \{x = (x_1, x_2, x_3) : x_3 = 0\}$, and $B$ is the rest of the boundary of $\Omega$ i.e. $B = \partial\Omega \setminus F$. The problem (4.11 - 4.13) describes small oscillations of an ideal fluid in a container $\Omega$. The fluid occupies the whole container $\Omega$. $B$ is the bottom of the container and $F$ is the fluid's free surface where the fluid can oscillate. We assume that the fluid is incompressible, inviscid and irrotational. Then $\nabla u_n(x)$ is the maximal amplitude of the velocity of the fluid at the point $x$. This problem has been widely studied see e.g. [108], [115].

## 4.4   Eigenvalue Estimates

Z.-Q. Chen and R. Song obtained some very interesting two-sided eigenvalue estimates [57]. They obtained their result for a very wide class of subordinate Markov processes. Here we only present their results for symmetric $\alpha$-stable processes. Their main result for these processes is the following (cf. Example 5.1 [57]).

**Theorem 4.7.** *Fix* $\alpha \in (0, 2)$ *and let* $D \subset \mathbb{R}^d$ *be an open bounded set. Let* $\{\lambda_n\}_{n=1}^\infty$ *be eigenvalues of the semigroup* $\{P_t^D\}_{t \geq 0}$ *and* $\{\mu_n\}_{n=1}^\infty$ *be eigenvalues of the semigroup* $\{\tilde{P}_t^D\}_{t \geq 0}$ *for the Brownian motion (in other words* $\{\mu_n\}_{n=1}^\infty$ *are eigenvalues for the Dirichlet Laplacian).*
*Then we have*
*i)*

$$\lambda_n \leq \mu_n^{\alpha/2};$$

*ii) if additionally* $D$ *is a convex bounded domain,*

$$\frac{1}{2}\mu_n^{\alpha/2} \leq \lambda_n \leq \mu_n^{\alpha/2}.$$

The method of the proof is very interesting. Two processes are considered. The first one is Brownian motion subordinated by the stable subordinator (the symmetric stable process ) killed by leaving a domain. The second one is Brownian motion killed by leaving a domain and then subordinated by the stable subordinator. The main result is obtained by comparing the quadratic forms of these two processes.

## 4.5  Generalized Isoperimetric Inequalities

**Theorem 4.8.** *Let us fix $\alpha \in (0, 2]$. Let $D \subset \mathbb{R}^d$ be a bounded domain, $D^*$ the ball in $\mathbb{R}^d$ with the same volume as $D$. Then we have*

$$\lambda_1(D^*) \leq \lambda_1(D).$$

This theorem states that among all sets of equal volume the smallest first eigenvalue has the ball. It is well known for many years.

*Proof.* (idea) It is well known that for any $x \in D$

$$\lambda_1(D) = - \lim_{t \to \infty} \frac{1}{t} \log(P^x(\tau_D > t)).$$

We can and do assume that $0 \in D$ and $D^*$ has center at 0. We have

$$P^0(\tau_D > t) = \lim_{m \to \infty} P^0(X_{\frac{t}{m}} \in D, X_{\frac{2t}{m}} \in D, \ldots, X_{\frac{mt}{m}} \in D,)$$

$$= \int_D \int_D \cdots \int_D p_{\frac{t}{m}}(x_1) p_{\frac{t}{m}}(x_2 - x_1) \ldots p_{\frac{t}{m}}(x_m - x_{m-1}) \, dx_1 \, dx_2 \ldots dx_m$$

$$\leq \int_{D^*} \int_{D^*} \cdots \int_{D^*} p^*_{\frac{t}{m}}(x_1) p^*_{\frac{t}{m}}(x_2 - x_1) \ldots p^*_{\frac{t}{m}}(x_m - x_{m-1}) \, dx_1 \, dx_2 \ldots dx_m$$

$$= \int_{D^*} \int_{D^*} \cdots \int_{D^*} p_{\frac{t}{m}}(x_1) p_{\frac{t}{m}}(x_2 - x_1) \ldots p_{\frac{t}{m}}(x_m - x_{m-1}) \, dx_1 \, dx_2 \ldots dx_m.$$

$f^*(x)$ denotes here a symmetric decreasing rearrangement of $f(x)$. Since $p_t(x)$ is radial and radially decreasing we have $p_t^*(x) = p_t(x)$.

It follows that

$$\lambda_1(D) = - \lim_{t \to \infty} \frac{1}{t} \log(P^0(\tau_D > t))$$

$$\geq - \lim_{t \to \infty} \frac{1}{t} \log(P^0(\tau_{D^*} > t))$$

$$= \lambda_1(D^*) \qquad \qquad \square$$

**The estimates of the spectral gap $\lambda_2 - \lambda_1$.**

Let $\alpha \in (0, 2]$ and $D \subset \mathbb{R}^d$ be a bounded domain. Let us recall that there exists $C(D, \alpha)$ such that for all $t \geq 1$ we have

$$e^{-(\lambda_2 - \lambda_1)t} \leq \sup_{x, y \in D} \left| \frac{p_D(t, x, y)}{e^{-\lambda_1 t} \varphi_1(x) \varphi_1(y)} - 1 \right| \leq C(D, \alpha) e^{-(\lambda_2 - \lambda_1)t}.$$

In other words the spectral gap $\lambda_2 - \lambda_1$ measures the rate how quickly $p_D(t, x, y)/(e^{-\lambda_1 t} \varphi_1(x) \varphi_1(y))$ tends to 1.

The spectral gap for Laplacian and Schrödinger operator has been investigated for many years. Motivated by problems concerning the behavior of free

Boson gases M. van den Berg made the following conjecture for all convex domains $D \subset \mathbb{R}^d$ and convex and nonnegatives $V : D \to [0, \infty)$

$$\lambda_2^V - \lambda_1^V > \frac{3\pi^2}{d_D^2}, \tag{4.14}$$

where $d_D = \mathrm{diam}(D)$ and $\lambda_1^V$, $\lambda_2^V$ are the first and the second eigenvalue for the following eigenvalue problem for Schrödinger operator on $D$

$$\begin{cases} \Delta\varphi_n(x) - V(x)\varphi_n(x) = -\lambda_n\varphi_n(x); & x \in D, \\ \varphi_n(x) = 0; & x \in \partial D. \end{cases}$$

The case $V \equiv 0$ corresponds to Dirichlet Laplacian.

Let us point out that for a rectangle $D = (0, a) \times (0, b)$, $a > b$ and $V \equiv 0$ we have

$$\varphi_1(x) = \sin\left(\frac{\pi x_1}{a}\right) \sin\left(\frac{\pi x_2}{b}\right), \quad \lambda_1 = \frac{\pi^2}{a^2} + \frac{\pi^2}{b^2},$$

$$\varphi_2(x) = \sin\left(\frac{2\pi x_1}{a}\right) \sin\left(\frac{\pi x_2}{b}\right), \quad \lambda_2 = \frac{4\pi^2}{a^2} + \frac{\pi^2}{b^2},$$

so the spectral gap equals $\lambda_2 - \lambda_1 = 3\pi^2/a^2$. By taking sufficiently long rectangles we can see that the estimate (4.14) is optimal.

Let us also note that for the conjecture (4.14) the assumption of convexity of $D$ is necessary. If we take domains $D$ with constant diameter which are not convex then the spectral gap can be arbitrarily small. It is sufficient to consider $V \equiv 0$ and the domain consisting of two disjoint balls joint together with sufficiently narrow corridor.

The first substantial progress on this conjecture was made in 1985 by I. M. Singer, B. Wong, S. T. Yau and S. S. T. Yau [141], who used a maximum principle technique to show that

$$\lambda_2^V - \lambda_1^V \geq \frac{\pi^2}{4d_D^2}.$$

Later Q. Yu i J. Q. Zhong in 1986 [157] showed that

$$\lambda_2^V - \lambda_1^V \geq \frac{\pi^2}{d_D^2}.$$

Next in 1993 J. Ling [119] obtained the strict inequality

$$\lambda_2^V - \lambda_1^V > \frac{\pi^2}{d_D^2}.$$

In 2000 B. Davis [68] and independently R. Bañuelos and P. Mendez [9] showed that if $D \subset \mathbb{R}^2$ is convex and double symmetric (that is symmetric according to both coordinate axes) and $V \equiv 0$ then conjecture (4.14) holds i.e.

$$\lambda_2^V - \lambda_1^V > \frac{3\pi^2}{d_D^2}.$$

Most (but not all) of the methods of estimating the spectral gap for Laplacian and Schrödinger operator uses the fact that the first eigenfunction on a convex domain with a convex potential is logconcave.

Unfortunately for $\alpha \in (0, 2)$ we do not know whether the first eigenfunction is logconcave on a convex domain. For $\alpha \in (0, 2)$ even the weaker property that the first eigenfunction is *unimodal* would be very interesting and helpful to estimate the spectral gap. Therefore we have the following open problem.

**Conjecture 4.9.** *Let $\alpha \in (0, 2)$ and $D \subset \mathbb{R}^d$ be a convex bounded domain. Then the first eigenfunction $\varphi_1$ for the semigroup $\{P_t^D\}_{t\geq0}$ is unimodal, that is $\varphi_1$ is unimodal along any line segment in $D$.*

Of course for $\alpha = 2$ the first eigenfunction $\varphi_1$ is unimodal because is logconcave.

Even though we do not know whether the first eigenfunction is unimodal for $\alpha \in (0, 2)$ some estimates of the spectral gap were obtained. First they were obtained for the Cauchy process ($\alpha = 1$) using the connection with the Steklov problem. Later these results were extended for all $\alpha \in (0, 2)$.

The result for the Cauchy process is the following [8].

**Theorem 4.10.** *Let $D \subset \mathbb{R}^2$ be a bounded convex domain which is symmetric relative to both coordinate axes. Assume that $[-a, a] \times [-b, b]$, $a \geq b > 0$ is the smallest rectangle (with sides parallel to the coordinate axes) containing $D$. Let $\{\lambda_n\}_{n=1}^{\infty}$ be the eigenvalues corresponding to the semigroup of the Cauchy process killed upon exiting $D$. We have*

$$\lambda_2 - \lambda_1 \geq \frac{Cb}{a^2},$$

*where $C = 10^{-7}$ is an absolute constant.*

The estimate is obtained by proving a new weighted Poincaré inequality and appealing to the connection between the eigenvalue problem for the Cauchy process and a mixed boundary value problem for the Laplacian in one dimension higher known as the mixed Steklov problem established in [7].

For $\alpha \in (0, 2)$ we have the following variational formula [73]

$$\lambda_2 - \lambda_1 = \inf_{f \in \mathcal{F}} \frac{\mathcal{A}_{d,-\alpha}}{2} \int_D \int_D \frac{(f(x) - f(y))^2}{|x - y|^{d+\alpha}} \varphi_1(x)\varphi_1(y) \, dx \, dy, \qquad (4.15)$$

where

$$\mathcal{F} = \{f \in L^2(D, \varphi_1^2) : \int_D f^2(x)\varphi_1^2(x) \, dx = 1, \quad \int_D f(x)\varphi_1^2(x) \, dx = 0\}.$$

This formula is quite easy to prove. It has not been published before although it follows from some other more general papers see e.g. [47].

Using this formula the following result has been shown [73].

**Theorem 4.11.** *Let $D \subset \mathbb{R}^2$ be a bounded convex domain which is symmetric relative to both coordinate axes. Assume that $[-a, a] \times [-b, b]$, $a \geq b$ is the smallest rectangle (with sides parallel to the coordinate axes) containing $D$. Then we have*

$$2A_{2,-\alpha}^{-1}(\lambda_2 - \lambda_1) \geq \frac{C\, b^{2-\alpha}}{a^2},$$

*where*

$$C = C(\alpha) = 10^{-9} 3^{\alpha-4} 2^{-2\alpha-1} \left( 4 + \frac{12\Gamma(2/\alpha)}{\alpha(2-\alpha)(1-2^{-\alpha})^{2/\alpha}} \right)^{-2}. \qquad (4.16)$$

For rectangles it was possible to obtain the following more precise estimates [73]. These estimates are sharp i.e. the upper and lower bound estimates have the same dependence on the length of the sides of the rectangle. Nevertheless, the numerical constants which appear in this theorem are far from being optimal.

**Theorem 4.12.** *Let $D = (-a, a) \times (-b, b)$, where $a \geq b$. Then*

*(a)We have*

$$2A_{2,-\alpha}^{-1}(\lambda_2 - \lambda_1) \leq 10^6 \cdot \begin{cases} \dfrac{2}{1-\alpha} \dfrac{b}{a^{1+\alpha}} & \text{for } \alpha < 1, \\[2ex] 2\log\left(1 + \dfrac{a}{b}\right) \dfrac{b}{a^2} & \text{for } \alpha = 1, \\[2ex] \left(\dfrac{1}{2-\alpha} + \dfrac{1}{\alpha-1}\right) \dfrac{b^{2-\alpha}}{a^2} & \text{for } \alpha > 1. \end{cases}$$

*(b)We have*

$$2A_{2,-\alpha}^{-1}(\lambda_2 - \lambda_1) \geq \begin{cases} \dfrac{b}{36 \cdot 2^{1+2\alpha} a^{1+\alpha}} & \text{for } \alpha < 1, \\[2ex] 10^{-9} \log\left(1 + \dfrac{a}{b}\right) \dfrac{b}{a^2} & \text{for } \alpha = 1, \\[2ex] \dfrac{1}{33 \cdot 13^{1+\alpha/2} \cdot 10^4} \dfrac{b^{2-\alpha}}{a^2} & \text{for } \alpha > 1. \end{cases}$$

**Remark 4.13.** *The inequality*

$$2A_{2,-\alpha}^{-1}(\lambda_2 - \lambda_1) \geq \frac{b}{36 \cdot 2^\alpha (a+b)^{1+\alpha}}$$

*holds for all $\alpha \in (0, 2)$.*

We have $2\mathcal{A}_{2,-\alpha}^{-1} = \alpha^{-2}2^{3-\alpha}\pi\Gamma^{-1}(\alpha/2)\Gamma(1-\alpha/2)$. *In particular we get for example* $\lambda_2 - \lambda_1 \geq \frac{8b}{10^4(a+b)^{3/2}}$ *for* $\alpha = 1/2$, $\lambda_2 - \lambda_1 \geq \frac{b}{10^3(a+b)^2}$ *for* $\alpha = 1$, $\lambda_2 - \lambda_1 \geq \frac{8b}{10^4(a+b)^{5/2}}$ *for* $\alpha = 3/2$.

For $\alpha \in (0,2)$ it is also possible to obtain some estimates for the spectral gap for all bounded open sets not necessarily convex or not even connected. M. Kwaśnicki [116] showed the following result.

**Theorem 4.14.** *Let* $\alpha \in (0,2)$ *and* $D \subset \mathbb{R}^d$ *be a bounded open set with inradius* 1. *Then*

    *i)*

$$\|\varphi_1\|_\infty \leq c_1, \quad c_1 = c_1(d,\alpha),$$

    *ii)*

$$\lambda_2 - \lambda_1 \geq \frac{c_2}{d_D^{d+\alpha}}, \quad c_2 = c_2(d,\alpha).$$

Let us notice that for convex double symmetric domains the estimates obtained by R. Bañuelos and T. Kulczycki are sharper.

*Proof.* (implication i) → ii)) Let us denote the right hand side of (4.15) by $\mathcal{E}(f,f)$. Let $f \in mathcalF$ which is equivalent to $\int_D f(x)\varphi_1^2(x)\,dx = 0$ and $\int_D f^2(x)\varphi_1^2(x)\,dx = 1$. We have

$$\mathcal{E}(f,f) \geq \frac{\mathcal{A}_{d,-\alpha}}{2d_D^{d+\alpha}} \int_D \int_D (f(x)-f(y))^2 \varphi_1(x)\varphi_1(y)\,dx\,dy$$

$$\geq \frac{\mathcal{A}_{d,-\alpha}}{2d_D^{d+\alpha}c_1^2} \int_D \int_D (f(x)-f(y))^2 \varphi_1^2(x)\varphi_1^2(y)\,dx\,dy$$

$$= \frac{\mathcal{A}_{d,-\alpha}}{2d_D^{d+\alpha}c_1^2} \left( \int_D f^2(x)\varphi_1^2(x)\,dx \int_D \varphi_1^2(y)\,dy \right.$$

$$\left. - 2\int_D f(x)\varphi_1^2(x)\,dx \int_D f(y)\varphi_1^2(y)\,dy + \int_D f^2(y)\varphi_1^2(y)\,dy \int_D \varphi_1^2(x)\,dx \right)$$

$$= \frac{\mathcal{A}_{d,-\alpha}}{2d_D^{d+\alpha}c_1^2} 2\int_D f^2(x)\varphi_1^2(x)\,dx$$

$$= \frac{c_2}{d_D^{d+\alpha}} \int_D f^2(x)\varphi_1^2(x)\,dx. \qquad\qquad \square$$

The constant $d+\alpha$ in the power of $\frac{1}{d_D}$ is optimal which can be seen from the following example. Let $D = B_1 \cup B_2$ where $B_1, B_2$ are two disjoint unit balls in $\mathbb{R}^d$. Then we have

$$\lambda_2 - \lambda_1 \leq \frac{\mathcal{E}(f,f)}{\int_D f^2\varphi_1^2} = \mathcal{E}(f,f) \leq \frac{c(d,\alpha)}{d_D^{d+\alpha}}.$$

# Chapter 5
# Potential Theory of Subordinate Brownian Motion

by R. Song and Z. Vondraček

## 5.1 Introduction

The materials covered in the second part of the book are based on several recent papers, primarily [132], [139], [148] and [146]. The main effort here was given to unify the exposition of those results, and in doing so we also eradicated the typos in these papers. Some new materials and generalizations are also included. Here is the outline of Chapter 5.

In Section 5.2 we recall some basic facts about subordinators and give a list of examples that will be useful later on. This list contains stable subordinators, relativistic stable subordinators, subordinators which are sums of stable subordinators and a drift, gamma subordinators, geometric stable subordinators, iterated geometric stable subordinators and Bessel subordinators. All of these subordinators belong to the class of special subordinators (even complete Bernstein subordinators). Special subordinators are important to our approach because they are precisely the ones whose potential measure restricted to $(0, \infty)$ has a decreasing density $u$. In fact, for all of the listed subordinators the potential measure has a decreasing density $u$. In the last part of the section we study asymptotic behaviors of the potential density $u$ and the Lévy density of subordinators by use of Karamata's and de Haan's Tauberian and monotone density theorems.

In Section 5.3 we derive asymptotic properties of the Green function and the jumping function of subordinate Brownian motion. These results follow from the technical Lemma 5.32 upon checking its conditions for particular subordinators. Of special interest is the order of singularities of the Green function near zero, starting from the Newtonian kernel at the one end, and singularities on the brink of integrability on the other end obtained for iterated geometric stable subordinators. The results for the asymptotic behavior of the jumping function are less complete, but are substituted by results on the decay at zero and at infinity. Finally, we discuss transition densities for symmetric geometric stable processes which exhibit unusual behavior on the diagonal for small (as well as large) times.

K. Bogdan et al., *Potential Analysis of Stable Processes and its Extensions*, Lecture Notes in Mathematics 1980, DOI 10.1007/978-3-642-02141-1_5, © Springer-Verlag Berlin Heidelberg 2009

The original motivation for deriving the results in Sections 5.2 and 5.3 was an attempt to obtain the Harnack inequality for subordinate Brownian motions with subordinators whose Laplace exponent $\phi(\lambda)$ has the asymptotic behavior at infinity of one of the following two forms: (i) $\phi(\lambda) \sim \lambda$, or (ii) logarithmic behavior at $\infty$. A typical example of the first case is the process $Y$ which is a sum of Brownian motion and an independent rotationally invariant $\alpha$-stable process. This situation was studied in [132]. A typical example of the second case is a geometric stable process – a subordinate Brownian motion via a geometric stable subordinator. In this case, $\phi(\lambda) \sim \log \lambda$ as $\lambda \to \infty$. This was studied in [139]. Section 5.4 contains an exposition of these results and some generalizations, and is partially based on the general approach to Harnack inequality from [145]. After obtaining some potential-theoretic results for a class of radial Lévy processes, we derive Krylov-Safonov-type estimates for the hitting probabilities involving capacities. Similar estimates involving Lebesgue measure were obtained in [145] based on the work of Bass and Levin [13]. These estimates are crucial in proving two types of Harnack inequalities for small balls - scale invariant ones, and the weak ones in which the constant might depend on the radius of a ball. In fact, we give a full proof of the Harnack inequality only for iterated geometric stable processes, and refer the reader to the original papers for the other cases.

Finally, in Section 5.5 we replace the underlying Brownian motion by the Brownian motion killed upon exiting a Lipschitz domain $D$. The resulting process is denoted by $X^D$. We are interested in the potential theory of the process $Y_t^D = X^D(S_t)$ where $S$ is a special subordinator with infinite Lévy measure or positive drift. Such questions were first studied for stable subordinators in [81], and the final solution in this case was given in [80]. The general case for special subordinators appeared in [148]. Surprisingly, it turns out that the potential theory of $Y^D$ is in a one-to-one and onto correspondence with the potential theory of $X^D$. More precisely, there is a bijection (realized by the potential operator of the subordinate process $Z_t^D = X^D(T_t)$ where $T$ is the subordinator conjugate to $S$) from the cone $\mathcal{S}(Y^D)$ of excessive functions of $Y^D$ onto the cone $\mathcal{S}(X^D)$ of excessive functions for $X^D$ which preserves nonnegative harmonic functions. This bijection makes it possible to essentially transfer the potential theory of $X^D$ to the potential theory of $Y^D$. In this way we obtain the Martin kernel and the Martin representation for $Y^D$ which immediately leads to a proof of the boundary Harnack principle for nonnegative harmonic functions of $Y^D$. In the case of a $C^{1,1}$ domain we obtain sharp bounds for the transition densities of the subordinate process $Y^D$.

The materials covered in this part of the book by no means include all that can be said about the potential theory of subordinate Brownian motions. One of the omissions is the Green function estimates for killed subordinate Brownian motions and the boundary Harnack inequality for the positive harmonic functions of subordinate Brownian motions. By using ideas from [53] or [134] one can easily extend the Green function estimates of [51] and [109] for killed symmetric stable processes to more general killed subordinate Brownian motions under certain conditions, and then use these estimates to extend arguments in [27] and [150] to establish the boundary Harnack inequality for general subordinate Brownian motions under certain conditions. In the case when the Laplace exponent $\phi$ is regularly varying at infinity, this is done in [105]. Another notable omission is the spectral theory for such processes together with implications to spectral theory of killed subordinate Brownian motion. We refer the reader to [57], [58] and [59]. Related to this is the general discussion on the exact difference between subordinate killed Brownian motions and the killed subordinate Brownian motions and its consequences. This was discussed in [144] and [143]. See also [87] and [149].

We end this introduction with few words on the notations. For functions $f$ and $g$ we write $f \sim g$ if the quotient $f/g$ converges to 1, and $f \asymp g$ if the quotient $f/g$ stays bounded between two positive constants.

## 5.2  Subordinators

### 5.2.1  Special Subordinators and Complete Bernstein Functions

Let $S = (S_t : t \geq 0)$ be a subordinator, that is, an increasing Lévy process taking values in $[0, \infty]$ with $S_0 = 0$. We remark that our subordinators are what some authors call killed subordinators. The Laplace transform of the law of $S_t$ is given by the formula

$$\mathbb{E}[\exp(-\lambda S_t)] = \exp(-t\phi(\lambda)), \quad \lambda > 0. \tag{5.1}$$

The function $\phi : (0, \infty) \to \mathbb{R}$ is called the Laplace exponent of $S$, and it can be written in the form

$$\phi(\lambda) = a + b\lambda + \int_0^\infty (1 - e^{-\lambda t})\, \mu(dt). \tag{5.2}$$

Here $a, b \geq 0$, and $\mu$ is a $\sigma$-finite measure on $(0, \infty)$ satisfying

$$\int_0^\infty (t \wedge 1)\, \mu(dt) < \infty. \tag{5.3}$$

The constant $a$ is called the killing rate, $b$ the drift, and $\mu$ the Lévy measure of the subordinator $S$. By using condition (5.3) above one can easily check that

$$\lim_{t \to 0} t\, \mu(t, \infty) = 0, \tag{5.4}$$

$$\int_0^1 \mu(t, \infty)\, dt < \infty. \tag{5.5}$$

For $t \geq 0$, let $\eta_t$ be the distribution of $S_t$. To be more precise, for a Borel set $A \subset [0, \infty)$, $\eta_t(A) = \mathbb{P}(S_t \in A)$. The family of measures $(\eta_t : t \geq 0)$ forms a convolution semigroup of measures on $[0, \infty)$. Clearly, the formula (5.1) reads $\exp(-t\phi(\lambda)) = \mathcal{L}\eta_t(\lambda)$, the Laplace transform of the measure $\eta_t$. We refer the reader to [18] for much more detailed exposition on subordinators.

Recall that a $C^\infty$ function $\phi : (0, \infty) \to [0, \infty)$ is called a Bernstein function if $(-1)^n D^n \phi \leq 0$ for every $n \in \mathbb{N}$. It is well known (see, e.g., [17]) that a function $\phi : (0, \infty) \to \mathbb{R}$ is a Bernstein function if and only if it has the representation given by (5.2).

We now introduce the concepts of special Bernstein functions and special subordinators.

**Definition 5.1.** *A Bernstein function $\phi$ is called a special Bernstein function if $\psi(\lambda) := \lambda/\phi(\lambda)$ is also a Bernstein function. A subordinator $S$ is called a special subordinator if its Laplace exponent is a special Bernstein function.*

We will call the function $\psi$ in the definition above the Bernstein function *conjugate* to $\phi$.

Special subordinators occur naturally in various situations. For instance, they appear as the ladder time process for a Lévy process which is not a compound Poisson process, see page 166 of [18]. Yet another situation in which they appear naturally is in connection with the exponential functional of subordinators (see [20]).

The most common examples of special Bernstein functions are complete Bernstein functions, also called operator monotone functions in some literature. A function $\phi : (0, \infty) \to \mathbb{R}$ is called a complete Bernstein function if there exists a Bernstein function $\eta$ such that

$$\phi(\lambda) = \lambda^2 \mathcal{L}\eta(\lambda), \quad \lambda > 0,$$

where $\mathcal{L}$ stands for the Laplace transform of the function $\eta$: $\mathcal{L}\eta(\lambda) = \int_0^\infty e^{-\lambda t} \eta(t)\, dt$. It is known (see, for instance, Remark 3.9.28 and Theorem 3.9.29 of [97]) that every complete Bernstein function is a Bernstein function and that the following three conditions are equivalent:

(i) $\phi$ is a complete Bernstein function;
(ii) $\psi(\lambda) := \lambda/\phi(\lambda)$ is a complete Bernstein function;
(iii) $\phi$ is a Bernstein function whose Lévy measure $\mu$ is given by

$$\mu(dt) = \int_0^\infty e^{-st} \gamma(ds) dt$$

where $\gamma$ is a measure on $(0, \infty)$ satisfying

$$\int_0^1 \frac{1}{s} \gamma(ds) + \int_1^\infty \frac{1}{s^2} \gamma(ds) < \infty.$$

The equivalence of (i) and (ii) says that every complete Bernstein function is a special Bernstein function. Note also that it follows from the condition (iii) above that being a complete Bernstein function only depends on the Lévy measure and that the Lévy measure $\mu(dt)$ of any complete Bernstein function has a completely monotone density. We also note that the tail $t \to \mu(t, \infty)$ of the Lévy measure $\mu$ is a completely monotone function. Indeed, by Fubini's theorem

$$\mu(x, \infty) = \int_x^\infty \int_0^\infty e^{-st} \gamma(ds)\, dt = \int_0^\infty e^{-xs} \frac{\gamma(ds)}{s}.$$

A similar argument shows that the converse is also true, namely, if the tail of the Lévy measure $\mu$ is a completely monotone function, then $\mu$ has a completely monotone density. The density of the Lévy measure with respect to the Lebesgue measure (when it exists) will be called the Lévy density.

The family of all complete Bernstein functions is a closed convex cone containing positive constants. The following properties of complete Bernstein functions are well known, see, for instance, [126]: (i) If $\phi$ is a nonzero complete Bernstein function, then so are $\phi(\lambda^{-1})^{-1}$ and $\lambda\phi(\lambda^{-1})$; (ii) if $\phi_1$ and $\phi_2$ are nonzero complete Bernstein functions and $\beta \in (0, 1)$, then $\phi_1^\beta(\lambda)\phi_2^{1-\beta}(\lambda)$ is also a complete Bernstein function; (iii) if $\phi_1$ and $\phi_2$ are nonzero complete Bernstein functions and $\beta \in (-1, 0) \cup (0, 1)$, then $(\phi_1^\beta(\lambda) + \phi_2^\beta(\lambda))^{1/\beta}$ is also a complete Bernstein function.

Most of the familiar Bernstein functions are complete Bernstein functions. The following are some examples of complete Bernstein functions ([97]): (i) $\lambda^\alpha, \alpha \in (0, 1]$; (ii) $(\lambda + 1)^\alpha - 1, \alpha \in (0, 1)$; (iii) $\log(1 + \lambda)$; (iv) $\frac{\lambda}{\lambda+1}$. The first family corresponds to $\alpha$-stable subordinators ($0 < \alpha < 1$) and a pure drift ($\alpha = 1$), the second family corresponds to relativistic $\alpha$-stable subordinators, the third Bernstein function corresponds to the gamma subordinator, and the fourth corresponds to the compound Poisson process with rate 1 and exponential jumps. An example of a Bernstein function which is not a complete Bernstein function is $1 - e^{-\lambda}$. One can also check that $1 - e^{-\lambda}$ is not a special Bernstein function as well. We refer the reader to [138] for an extensive treatment of complete Bernstein functions.

The potential measure of the subordinator $S$ is defined by

$$U(A) = \mathbb{E}\int_0^\infty 1_{(S_t \in A)}\, dt = \int_0^\infty \eta_t(A)\, dt, \quad A \subset [0, \infty). \qquad (5.6)$$

Note that $U(A)$ is the expected time the subordinator $S$ spends in the set $A$. The Laplace transform of the measure $U$ is given by

$$\mathcal{L}U(\lambda) = \int_0^\infty e^{-\lambda t}\, dU(t) = \mathbb{E}\int_0^\infty \exp(-\lambda S_t)\, dt = \frac{1}{\phi(\lambda)}. \qquad (5.7)$$

We are going to derive a characterization of special subordinators in terms of their potential measures. Roughly, a subordinator $S$ is special if and only if its potential measure $U$ restricted to $(0, \infty)$ has a decreasing density. To be more precise, let $S$ be a special subordinator with the Laplace exponent $\phi$ given by

$$\phi(\lambda) = a + b\lambda + \int_0^\infty (1 - e^{-\lambda t})\, \mu(dt)\,.$$

Then

$$\lim_{\lambda \to 0} \frac{\lambda}{\phi(\lambda)} = \begin{cases} 0\,, & a > 0, \\ \frac{1}{b + \int_0^\infty t\,\mu(dt)}\,, & a = 0, \end{cases}$$

$$\lim_{\lambda \to \infty} \frac{1}{\phi(\lambda)} = \begin{cases} 0\,, & b > 0 \text{ or } \mu(0, \infty) = \infty, \\ \frac{1}{a + \mu(0, \infty)}\,, & b = 0 \text{ and } \mu(0, \infty) < \infty\,. \end{cases}$$

Since $\lambda/\phi(\lambda)$ is a Bernstein function, we must have

$$\frac{\lambda}{\phi(\lambda)} = \tilde{a} + \tilde{b}\lambda + \int_0^\infty (1 - e^{-\lambda t})\, \nu(dt) \tag{5.8}$$

for some Lévy measure $\nu$, and

$$\tilde{a} = \begin{cases} 0\,, & a > 0, \\ \frac{1}{b + \int_0^\infty t\,\mu(dt)}\,, & a = 0, \end{cases} \tag{5.9}$$

$$\tilde{b} = \begin{cases} 0\,, & b > 0 \text{ or } \mu(0, \infty) = \infty, \\ \frac{1}{a + \mu(0, \infty)}\,, & b = 0 \text{ and } \mu(0, \infty) < \infty\,. \end{cases} \tag{5.10}$$

Equivalently,

$$\frac{1}{\phi(\lambda)} = \tilde{b} + \int_0^\infty e^{-\lambda t} \tilde{\Pi}(t)\, dt \tag{5.11}$$

with

$$\tilde{\Pi}(t) = \tilde{a} + \nu(t, \infty)\,, \quad t > 0\,.$$

Let $\tau(dt) := \tilde{b}\epsilon_0(dt) + \tilde{\Pi}(t)\, dt$. Then the right-hand side in (5.11) is the Laplace transform of the measure $\tau$. Since $1/\phi(\lambda) = \mathcal{L}U(\lambda)$, the Laplace transform of the potential measure $U$ of $S$, we have that $\mathcal{L}U(\lambda) = \mathcal{L}\tau(\lambda)$. Therefore,

$$U(dt) = \tilde{b}\epsilon_0(dt) + u(t)\, dt\,,$$

with a decreasing function $u(t) = \tilde{\Pi}(t)$.

Conversely, suppose that $S$ is a subordinator with potential measure given by

$$U(dt) = c\epsilon_0(dt) + u(t)\, dt\,,$$

for some $c \geq 0$ and some decreasing function $u : (0, \infty) \to (0, \infty)$ satisfying $\int_0^1 u(t)\, dt < \infty$. Then

$$\frac{1}{\phi(\lambda)} = \mathcal{L}U(\lambda) = c + \int_0^\infty e^{-\lambda t} u(t)\, dt\,.$$

It follows that

$$\frac{\lambda}{\phi(\lambda)} = c\lambda + \int_0^\infty u(t)\, d(1 - e^{-\lambda t})$$

$$= c\lambda + u(t)(1 - e^{-\lambda t}) \mid_0^\infty - \int_0^\infty (1 - e^{-\lambda t})\, u(dt)$$

$$= c\lambda + u(\infty) + \int_0^\infty (1 - e^{-\lambda t})\, \gamma(dt)\,, \tag{5.12}$$

with $\gamma(dt) = -u(dt)$. In the last equality we used that $\lim_{t \to 0} u(t)(1 - e^{-\lambda t}) = 0$. This is a consequence of the assumption $\int_0^1 u(t)\, dt < \infty$. It is easy to check, by using the same integrability condition on $u$, that $\int_0^\infty (1 \wedge t)\, \gamma(dt) < \infty$, so that $\gamma$ is a Lévy measure. Therefore, $\lambda/\phi(\lambda)$ is a Bernstein function, implying that $S$ is a special subordinator.

In this way we have proved the following

**Theorem 5.1.** *Let $S$ be a subordinator with the potential measure $U$. Then $S$ is special if and only if*

$$U(dt) = c\epsilon_0(dt) + u(t)\, dt$$

*for some $c \geq 0$ and some decreasing function $u : (0, \infty) \to (0, \infty)$ satisfying $\int_0^1 u(t)\, dt < \infty$.*

**Remark 5.2.** *The above result appeared in [19] as Corollaries 1 and 2 and was possibly known even before. The above presentation is taken from [148]. In case $c = 0$, we will call $u$ the potential density of the subordinator $S$ (or of the Laplace exponent $\phi$).*

**Corollary 5.3.** *Let $S$ be a subordinator with the Laplace exponent $\phi$ and the potential measure $U$. Then $\phi$ is a complete Bernstein function if and only if $U$ restricted to $(0, \infty)$ has a completely monotone density $u$.*

**Proof.** Note that from the proof of Theorem 5.1 we have the explicit form of the density $u$: $u(t) = \tilde{\Pi}(t)$ where $\tilde{\Pi}(t) = \tilde{a} + \nu(t, \infty)$. Here $\nu$ is the Lévy measure of $\lambda/\phi(\lambda)$. If $\phi$ is complete Bernstein, then $\lambda/\phi(\lambda)$ is complete Bernstein, and hence it follows from the property (iii) of complete Bernstein function that $u(t) = \tilde{a} + \nu(t, \infty)$ is a completely monotone function. Conversely, if $u$ is completely monotone, then clearly the tail $t \to \nu(t, \infty)$ is completely monotone, which implies that $\lambda/\phi(\lambda)$ is complete Bernstein. Therefore, $\phi$ is also a complete Bernstein function. □

Note that by comparing expressions (5.8) and (5.12) for $\lambda/\phi(\lambda)$, and by using formulae (5.9) and (5.10), it immediately follows that

$$c = \tilde{b} = \begin{cases} 0, & b > 0 \text{ or } \mu(0,\infty) = \infty, \\ \frac{1}{a+\mu(0,\infty)}, & b = 0 \text{ and } \mu(0,\infty) < \infty, \end{cases}$$

$$u(\infty) = \tilde{a} = \begin{cases} 0, & a > 0, \\ \frac{1}{b+\int_0^\infty t\,\mu(dt)}, & a = 0, \end{cases}$$

$$u(t) = \tilde{a} + \nu(t,\infty).$$

In particular, it cannot happen that both $a$ and $\tilde{a}$ are positive, and similarly, that both $b$ and $\tilde{b}$ are positive. Moreover, it is clear from the definition of $\tilde{b}$ that $\tilde{b} > 0$ if and only if $b = 0$ and $\mu(0,\infty) < \infty$.

We record now some consequences of Theorem 5.1 and the formulae above.

**Corollary 5.4.** *Suppose that* $S = (S_t : t \geq 0)$ *is a subordinator whose Laplace exponent*

$$\phi(\lambda) = a + b\lambda + \int_0^\infty (1 - e^{-\lambda t})\,\mu(dt)$$

*is a special Bernstein function with* $b > 0$ *or* $\mu(0,\infty) = \infty$. *Then the potential measure* $U$ *of* $S$ *has a decreasing density* $u$ *satisfying*

$$\lim_{t\to 0} t\,u(t) = 0, \tag{5.13}$$

$$\lim_{t\to 0} \int_0^t s\,du(s) = 0. \tag{5.14}$$

**Proof.** The formulae follow immediately from $u(t) = \tilde{a} + \nu(t,\infty)$ and (5.4)–(5.5) applied to $\nu$. □

**Corollary 5.5.** *Suppose that* $S = (S_t : t \geq 0)$ *is a special subordinator with the Laplace exponent given by*

$$\phi(\lambda) = a + \int_0^\infty (1 - e^{-\lambda t})\,\mu(dt)$$

*where* $\mu$ *satisfies* $\mu(0,\infty) = \infty$. *Then*

$$\psi(\lambda) := \frac{\lambda}{\phi(\lambda)} = \tilde{a} + \int_0^\infty (1 - e^{-\lambda t})\,\nu(dt) \tag{5.15}$$

*where the Lévy measure* $\nu$ *satisfies* $\nu(0,\infty) = \infty$.

*Let $T$ be the subordinator with the Laplace exponent $\psi$. If $u$ and $v$ denote the potential density of $S$ and $T$ respectively, then*

$$v(t) = a + \mu(t, \infty). \tag{5.16}$$

*In particular, $a = v(\infty)$ and $\tilde{a} = u(\infty)$. Moreover, $a$ and $\tilde{a}$ cannot be both positive.*

Assume that $\phi$ is a special Bernstein function with the representation (5.2) where $b > 0$ or $\mu(0, \infty) = \infty$. Let $S$ be a subordinator with the Laplace exponent $\phi$, and let $U$ denote its potential measure. By Corollary 5.4, $U$ has a decreasing density $u : (0, \infty) \to (0, \infty)$. Let $T$ be a subordinator with the Laplace exponent $\psi(\lambda) = \lambda/\phi(\lambda)$ and let $V$ denote its potential measure. Then $V(dt) = b\epsilon_0(dt) + v(t)\,dt$ where $v : (0, \infty) \to (0, \infty)$ is a decreasing function. If $b > 0$, the potential measure $V$ has an atom at zero, and hence the subordinator $T$ is a compound Poisson process (this can be also seen as follows: since $b > 0$, we have $u(0+) < \infty$, and hence $\nu(0, \infty) = u(0+) - \tilde{a} < \infty$). Note that in case $b > 0$, the Lévy measure $\mu$ can be finite. If $b = 0$, we require that $\mu(0, \infty) = \infty$, and then, by Corollary 5.5, $\psi(\lambda) = \lambda/\phi(\lambda)$ has the same form as $\phi$, namely $\tilde{b} = 0$ and $\nu(0, \infty) = \infty$. In this case, subordinators $S$ and $T$ play symmetric roles.

The following result will be crucial for the developments in Section 5.5 of this book.

**Theorem 5.6.** *Let $\phi$ be a special Bernstein function with representation (5.2) satisfying $b > 0$ or $\mu(0, \infty) = \infty$. Then*

$$b\,u(t) + \int_0^t u(s)v(t-s)\,ds = b\,u(t) + \int_0^t v(s)u(t-s)\,ds = 1, \quad t > 0. \tag{5.17}$$

**Proof.** Since for all $\lambda > 0$ we have

$$\frac{1}{\phi(\lambda)} = \mathcal{L}u(\lambda), \qquad \frac{\phi(\lambda)}{\lambda} = b + \mathcal{L}v(\lambda),$$

after multiplying we get

$$\frac{1}{\lambda} = b\mathcal{L}u(\lambda) + \mathcal{L}u(\lambda)\mathcal{L}v(\lambda)$$
$$= b\mathcal{L}u(\lambda) + \mathcal{L}(u * v)(\lambda).$$

Inverting this equality gives

$$1 = b\,u(t) + \int_0^t u(s)v(t-s)\,ds, \quad t > 0.$$

$\square$

Theorem 5.6 has an amusing consequence related to the first passage of the subordinator $S$. Let $\sigma_t = \inf\{s > 0 : S_s > t\}$ be the first passage time across the level $t > 0$. By the first passage formula (see, e.g., [18], p.76), we have

$$\mathbb{P}(S_{\sigma_{t-}} \in ds, S_{\sigma_t} \in dx) = u(s)\mu(x - s)\, ds\, dx\,,$$

for $0 \le s \le t$, and $x > t$. Since $\mu(x, \infty) = v(x)$, by use of Fubini's theorem this implies

$$\mathbb{P}(S_{\sigma_t} > t) = \int_t^\infty \int_0^t u(s)\mu(x - s)\, ds\, dx = \int_0^t u(s) \int_t^\infty \mu(x - s)\, dx\, ds$$
$$= \int_0^t u(s)\mu(t - s, \infty)\, ds = \int_0^t u(s)v(t - s)\, ds\,.$$

Since $\mathbb{P}(S_{\sigma_t} \ge t) = 1$, by comparing with (5.17) we see that $\mathbb{P}(S_{\sigma_t} = t) = b\, u(t)$. This provides a simple proof in case of special subordinators of the well-known fact true for general subordinators (see [18], pp.77–79).

In the sequel we will also need the following result on potential density that is valid for subordinators that are not necessarily special.

**Proposition 5.7.** *Let $S = (S_t : t \ge 0)$ be a subordinator with drift $b > 0$. Then its potential measure $U$ has a density $u$ continuous on $(0, \infty)$ satisfying $u(0+) = 1/b$ and $u(t) \le u(0+)$ for every $t > 0$.*

**Proof.** For the proof of existence of continuous $u$ and the fact that $u(0+) = 1/b$ see, e.g., [18], p.79. That $u(t) \le u(0+)$ for every $t > 0$ follows from the subadditivity of the function $t \mapsto U([0,t])$ (see, e.g., [132]). □

### 5.2.2 Examples of Subordinators

In this subsection we give a list of subordinators that will be relevant in the sequel and describe some of their properties.

**Example 5.8. (Stable subordinators)** Our first example covers the family of well-known stable subordinators. For $0 < \alpha < 2$, let $\phi(\lambda) = \lambda^{\alpha/2}$. By integration

$$\lambda^{\alpha/2} = \frac{\alpha/2}{\Gamma(1 - \alpha/2)} \int_0^\infty (1 - e^{-\lambda t})\, t^{-1-\alpha/2}\, dt\,,$$

i.e, the Lévy measure $\mu(dt)$ of $\phi$ has a density given by $(\alpha/2)/\Gamma(1 - \alpha/2)$ $t^{-1-\alpha/2}$. Since $t^{-1-\alpha/2} = \int_0^\infty e^{-ts} s^{\alpha/2}/\Gamma(1 + \alpha/2)\, ds$, it follows that $\phi$ is a complete Bernstein function. The tail of the Lévy measure $\mu$ is equal to

$$\mu(t, \infty) = \frac{t^{-\alpha/2}}{\Gamma(1 - \alpha/2)}\,.$$

The conjugate Bernstein function is $\psi(\lambda) = \lambda^{1-\alpha/2}$, hence its tail is $\nu(t, \infty) = t^{\alpha/2-1}/\Gamma(\alpha/2)$. This shows that the potential density of $\phi(\lambda) = \lambda^{\alpha/2}$ is equal to

$$u(t) = \frac{t^{\alpha/2-1}}{\Gamma(\alpha/2)}.$$

The subordinator $S$ corresponding to $\phi$ is called an $\alpha/2$-*stable subordinator*.

It is known that the distribution $\eta_1(ds)$ of the $\alpha/2$-stable subordinator has a density $\eta_1(s)$ with respect to the Lebesgue measure. Moreover, by [142],

$$\eta_1(s) \sim 2\pi\Gamma\left(1 + \frac{\alpha}{2}\right)\sin\left(\frac{\alpha\pi}{4}\right) s^{-1-\alpha/2}, \quad s \to \infty, \tag{5.18}$$

and

$$\eta_1(s) \le c(1 \wedge s^{-1-\alpha/2}), \quad s > 0, \tag{5.19}$$

for some positive constant $c > 0$.

**Example 5.9. (Relativistic stable subordinators)** For $0 < \alpha < 2$ and $m > 0$, let $\phi(\lambda) = (\lambda + m^{2/\alpha})^{\alpha/2} - m$. By integration

$$(\lambda + m^{2/\alpha})^{\alpha/2} - m = \frac{\alpha/2}{\Gamma(1-\alpha/2)}\int_0^\infty (1 - e^{-\lambda t})\, e^{-m^{2/\alpha}t}\, t^{-1-\alpha/2}\, dt,$$

i.e., the Lévy measure $\mu(dt)$ of $\phi$ has a density given by $(\alpha/2)/\Gamma(1-\alpha/2)$ $e^{-m^{2/\alpha}t}t^{-1-\alpha/2}$. This Bernstein function appeared in [118] in the study of the stability of relativistic matter, and so we call the corresponding subordinator $S$ a *relativistic $\alpha/2$-stable subordinator*. Since the Lévy density of $\phi$ is completely monotone, we know that $\phi$ is a complete Bernstein function. The explicit form of potential density $u$ of $S$ can be computed as follows (see [91] for this calculation): For $\gamma, \beta > 0$ let

$$E_{\gamma,\beta}(t) = \sum_{n=0}^\infty \frac{t^n}{\Gamma(\beta + \gamma n)}, \quad t > 0,$$

be the two parameter Mittag-Leffler function. By integrating term by term it follows that

$$\int_0^\infty e^{-\lambda t} e^{-m^{2/\alpha}t}\, t^{-1+\alpha/2} E_{\alpha/2,\alpha/2}(mt^{\alpha/2})\, dt = \frac{1}{(\lambda + m^{2/\alpha})^{\alpha/2} - m}.$$

Therefore,

$$u(t) = e^{-m^{2/\alpha}t}\, t^{-1+\alpha/2} E_{\alpha/2,\alpha/2}(mt^{\alpha/2}).$$

The subordinator $\tilde{S}$ corresponding to the complete Bernstein function $m + \phi(\lambda) = (\lambda + m^{2/\alpha})^{\alpha/2}$ is obtained by killing $S$ at an independent exponential time with parameter $m$. By checking tables of Laplace transforms ([74]) we see that

$$\frac{1}{m + \phi(\lambda)} = \int_0^\infty e^{-\lambda t} \frac{1}{\Gamma(\alpha/2)} e^{-m^{2/\alpha}t} t^{-1+\alpha/2} \, dt \,,$$

implying that the potential measure $\tilde{U}$ of the subordinator $\tilde{S}$ has the density $\tilde{u}$ given by

$$\tilde{u}(t) = \frac{1}{\Gamma(\alpha/2)} e^{-m^{2/\alpha}t} t^{-1+\alpha/2} \,.$$

**Example 5.10. (Gamma subordinator)** Let $\phi(\lambda) = \log(1 + \lambda)$. By use of Frullani's integral it follows that

$$\log(1 + \lambda) = \int_0^\infty (1 - e^{-\lambda t}) \frac{e^{-t}}{t} \, dt \,,$$

i.e., the Lévy measure of $\phi$ has a density given by $e^{-t}/t$. Note that $e^{-t}/t = \int_0^\infty e^{-st} 1_{(1,\infty)}(s) \, ds$, implying that the density of the Lévy measure $\mu$ is completely monotone. Therefore, $\phi$ is a complete Bernstein function. The corresponding subordinator $S$ is called a *gamma subordinator*. The explicit form of the potential density $u$ is not known. In the next section we will derive the asymptotic behavior of $u$ at 0 and at $+\infty$. On the other hand, the distribution $\eta_t(ds)$, $t > 0$, is well known and given by

$$\eta_t(ds) = \frac{1}{\Gamma(t)} s^{t-1} e^{-s} \, ds \,, \quad s > 0 \,. \tag{5.20}$$

Before proceeding to the next two examples, let us briefly discuss composition of subordinators. Suppose that $S^1 = (S_t^1 : t \geq 0)$ and $S^2 = (S_t^2 : t \geq 0)$ are two independent subordinators with Laplace exponents $\phi^1$, respectively $\phi^2$, and convolution semigroups $(\eta_t^1 : t \geq 0)$, respectively $(\eta_t^2 : t \geq 0)$. Define the new process $S = (S_t : t \geq 0)$ by $S_t = S^1(S_t^2)$, subordination of $S^1$ by $S^2$. Subordinating a Lévy process by an independent subordinator always yields a Lévy process (e.g. [135], p. 197). Hence, $S$ is another subordinator. The distribution $\eta_t$ of $S_t$ is given by

$$\eta_t(ds) = \int_0^\infty \eta_t^2(du) \eta_u^1(ds) \,. \tag{5.21}$$

Therefore, for any $\lambda > 0$,

$$\int_0^\infty e^{-\lambda s} \eta_t(ds) = \int_0^\infty e^{-\lambda s} \int_0^\infty \eta_t^2(du) \eta_u^1(ds)$$
$$= \int_0^\infty \eta_t^2(du) \int_0^\infty e^{-\lambda s} \eta_u^1(ds)$$
$$= \int_0^\infty \eta_t^2(du) e^{-u\phi^1(\lambda)} = \phi^2(\phi^1(\lambda)) \,,$$

showing that the Laplace exponent $\phi$ of $S$ is given by $\phi(\lambda) = \phi^2(\phi^1(\lambda))$.

**Example 5.11. (Geometric stable subordinators)** For $0 < \alpha < 2$, let $\phi(\lambda) = \log(1 + \lambda^{\alpha/2})$. Since $\phi$ is a composition of the complete Bernstein functions from Examples 5.8 and 5.10, it is itself a complete Bernstein function. The corresponding subordinator $S$ is called a *geometric $\alpha/2$-stable subordinator*. Note that this subordinator may be obtained by subordinating an $\alpha/2$-stable subordinator by a gamma subordinator. The concept of geometric stable distributions was first introduced in [106]. We will now compute the Lévy measure $\mu$ of $S$. Define

$$E_{\alpha/2}(t) := \sum_{n=0}^{\infty} (-1)^n \frac{t^{n\alpha/2}}{\Gamma(1 + n\alpha/2)}, \quad t > 0.$$

By checking tables of Laplace transforms (or by computing term by term), we see that

$$\int_0^{\infty} e^{-\lambda t} E_{\alpha/2}(t)\, dt = \frac{1}{\lambda(1 + \lambda^{-\alpha/2})} = \frac{\lambda^{\alpha/2 - 1}}{1 + \lambda^{\alpha/2}}. \tag{5.22}$$

Further, since $\phi(0+) = 0$ and $\lim_{\lambda \to \infty} \phi(\lambda)/\lambda = 0$, we have that $\phi(\lambda) = \int_0^{\infty}(1 - e^{-\lambda t})\,\mu(dt)$. By differentiating this expression for $\phi$ and the explicit form of $\phi$ we obtain that

$$\phi'(\lambda) = \int_0^{\infty} t e^{-\lambda t}\,\mu(dt) = \frac{\alpha}{2} \frac{\lambda^{\alpha/2 - 1}}{1 + \lambda^{\alpha/2}}. \tag{5.23}$$

By comparing (5.22) and (5.23) we see that the Lévy measure $\mu(dt)$ has a density given by

$$\mu(t) = \frac{\alpha}{2} \frac{E_{\alpha/2}(t)}{t}. \tag{5.24}$$

The explicit form of the potential density $u$ is not known. In the next section we will derive the asymptotic behavior of $u$ at $0+$ and at $\infty$.

We will now show that the distribution function of $S_1$ is given by

$$F(s) = 1 - E_{\alpha/2}(s) = \sum_{n=1}^{\infty} (-1)^{n-1} \frac{s^{n\alpha/2}}{\Gamma(1 + n\alpha/2)}, \quad s > 0. \tag{5.25}$$

Indeed, for $\lambda > 0$,

$$\mathcal{L}F(\lambda) = \int_0^{\infty} e^{-\lambda t} F(dt) = \lambda \int_0^{\infty} e^{-\lambda t}(1 - E_{\alpha/2}(t))\, dt$$

$$= \lambda \left( \frac{1}{\lambda} - \frac{\lambda^{\alpha/2 - 1}}{1 + \lambda^{\alpha/2}} \right) = \exp\{-\log(1 + \lambda^{\alpha/2})\}.$$

Since the function $\lambda \mapsto 1 + \lambda^{\alpha/2}$ is a complete Bernstein function, its reciprocal function, $\lambda \mapsto 1/(1 + \lambda^{\alpha/2})$ is a Stieltjes function (see [97] for more details about Stieltjes functions). Moreover, since $\lim_{t \to \infty} 1/(1 + \lambda^{\alpha/2}) = 0$, it follows that there exists a measure $\sigma$ on $(0, \infty)$ such that

$$\frac{1}{1 + \lambda^{\alpha/2}} = \mathcal{L}(\mathcal{L}\sigma)(\lambda).$$

But this means that the function $F$ has a completely monotone density $f$ given by $f(t) = \mathcal{L}\sigma(t)$. It is shown in [127] that the distribution function of $S_t$, $t > 0$, is equal to

$$\sum_{n=1}^{\infty} (-1)^{n-1} \frac{\Gamma(t + n - 1)s^{(t+n-1)\alpha/2}}{\Gamma(t)(n-1)!\Gamma(1 + (t + n - 1)\alpha/2)}.$$

Note that the case of the gamma subordinator may be subsumed under the case of geometric $\alpha/2$-stable subordinator by taking $\alpha = 2$ in the definition.

**Example 5.12. (Iterated geometric stable subordinators)** Let $0 < \alpha \leq 2$. Define,

$$\phi^{(1)}(\lambda) = \phi(\lambda) = \log(1 + \lambda^{\alpha/2}), \quad \phi^{(n)}(\lambda) = \phi(\phi^{(n-1)}(\lambda)), \quad n \geq 2.$$

Since $\phi^{(n)}$ is a complete Bernstein function, we have that $\phi^{(n)}(\lambda) = \int_0^{\infty} (1 - e^{-\lambda t})\mu^{(n)}(t)\,dt$ for a completely monotone Lévy density $\mu^{(n)}(t)$. The exact form of this density is not known.

Let $S^{(n)} = (S_t^{(n)} : t \geq 0)$ be the corresponding (iterated) subordinator, and let $U^{(n)}$ denote the potential measure of $S^{(n)}$. Since $\phi^{(n)}$ is a complete Bernstein function, $U^{(n)}$ admits a completely monotone density $u^{(n)}$. The explicit form of the potential density $u^{(n)}$ is not known. In the next section we will derive the asymptotic behavior of $u$ at 0 and at $+\infty$.

**Example 5.13. (Stable subordinators with drifts)** For $0 < \alpha < 2$ and $b > 0$, let $\phi(\lambda) = b\lambda + \lambda^{\alpha/2}$. Since $\lambda \mapsto \lambda^{\alpha/2}$ is complete Bernstein, it follows that $\phi$ is also a complete Bernstein function. The corresponding subordinator $S = (S_t : t \geq 0)$ is a sum of the pure drift subordinator $t \mapsto bt$ and the $\alpha/2$-stable subordinator. Its Lévy measure is the same as the Lévy measure of the $\alpha/2$-stable subordinator. In order to compute the potential density $u$ of the subordinator $S$, we first note that, similarly as in (5.22),

$$\int_0^{\infty} e^{-\lambda t} \frac{1}{b} E_{1-\alpha/2}(b^{-2/(2-\alpha)}t)\,dt = \frac{1}{b\lambda + \lambda^{\alpha/2}} = \frac{1}{\phi(\lambda)}.$$

Therefore, $u(t) = \frac{1}{b} E_{1-\alpha/2}(b^{-2/(2-\alpha)}t)$ for $t > 0$.

**Example 5.14.** (**Bessel subordinators**) The two subordinators in this example are taken from [122]. The Bessel subordinator $S_I = (S_I(t) : t \geq 0)$ is a subordinator with no drift, no killing and Lévy density

$$\mu_I(t) = \frac{1}{t} I_0(t) e^{-t},$$

where for any real number $\nu$, $I_\nu$ is the modified Bessel function. Since $\mu_I$ is the Laplace transform of the function $\gamma(t) = \int_0^t g(s)\, ds$ with

$$g(s) = \begin{cases} \pi^{-1}(2s - s^2)^{-1/2}, & s \in (0, 2), \\ 0, & s \geq 2, \end{cases}$$

the Laplace exponent of $S_I$ is a complete Bernstein function. The Laplace exponent of $S_I$ is given by

$$\phi_I(\lambda) = \log((1 + \lambda) + \sqrt{(1 + \lambda)^2 - 1}).$$

For any $t > 0$, the density of $S_I(t)$ is given by

$$f_t(x) = \frac{t}{x} I_t(x) e^{-x}.$$

The Bessel subordinator $S_K = (S_K(t) : t \geq 0)$ is a subordinator with no drift, no killing and Lévy density

$$\mu_K(t) = \frac{1}{t} K_0(t) e^{-t},$$

where for any real number $\nu$, $K_\nu$ is the modified Bessel function. Since $\mu_K$ is the Laplace transform of the function

$$\gamma(t) = \begin{cases} 0, & t \in (0, 2], \\ \log(t - 1 + \sqrt{(t-1)^2 + 1}), & t > 2, \end{cases}$$

the Laplace exponent of $S_K$ is a complete Bernstein function. The Laplace exponent of $S_K$ is given by

$$\phi_K(\lambda) = \frac{1}{2} \left( \log((1 + \lambda) + \sqrt{(1 + \lambda)^2 - 1}) \right)^2.$$

For any $t > 0$, the density of $S_K(t)$ is given by

$$f_t(x) = \sqrt{\frac{2\pi}{t}}\, \vartheta_x(\frac{1}{t}) \frac{e^{-x}}{x},$$

where

$$\vartheta_v(t) = \frac{v}{\sqrt{2\pi^3 t}} \int_0^\infty \exp(\frac{\pi^2 - \xi^2}{2t}) \exp(-v\cosh(\xi)) \sinh(\xi) \sin(\frac{\pi\xi}{t}) d\xi.$$

**Example 5.15.** For any $\alpha \in (0,2)$ and $\beta \in (0,2-\alpha)$, it follows from the properties of complete Bernstein functions that

$$\phi(\lambda) = \lambda^{\alpha/2}(\log(1+\lambda))^{\beta/2}$$

is a complete Bernstein function.

**Example 5.16.** For any $\alpha \in (0,2)$ and $\beta \in (0,\alpha)$, it follows from the properties of complete Bernstein functions that

$$\phi(\lambda) = \lambda^{\alpha/2}(\log(1+\lambda))^{-\beta/2}$$

is a complete Bernstein function.

### 5.2.3 Asymptotic Behavior of the Potential, Lévy and Transition Densities

Recall the formula (5.7) relating the Laplace exponent $\phi$ of the subordinator $S$ with the Laplace transform of its potential measure $U$. In the case $U$ has a density $u$, this formula reads

$$\mathcal{L}u(\lambda) = \int_0^\infty e^{-\lambda t} u(t)\, dt = \frac{1}{\phi(\lambda)}.$$

The asymptotic behavior of $\phi$ at $\infty$ (resp. at 0) determines, by use of Tauberian and the monotone density theorems, the asymptotic behavior of the potential density $u$ at 0 (resp. at $\infty$). We first recall Karamata's version of these theorems from [21].

**Theorem 5.17.** *(a) (Karamata's Tauberian theorem) Let $U : (0,\infty) \to (0,\infty)$ be an increasing function. If $\ell$ is slowly varying at $\infty$ (resp. at 0+), $\rho \geq 0$, the following are equivalent:*

*(i) As $t \to \infty$ (resp. $t \to 0+$)*

$$U(t) \sim \frac{t^\rho \ell(t)}{\Gamma(1+\rho)}.$$

*(ii) As $\lambda \to 0$ (resp. $\lambda \to \infty$)*

$$\mathcal{L}U(\lambda) \sim \lambda^{-\rho} \ell(1/\lambda).$$

*(b)(Karamata's monotone density theorem) If additionally $U(dx) = u(x) dx$, where $u$ is monotone and nonnegative, and $\rho > 0$, then (i) and (ii) are equivalent to:*

*(iii) As $t \to \infty$ (resp. $t \to 0+$)*

$$u(t) \sim \frac{\rho t^{\rho-1}\ell(t)}{\Gamma(1+\rho)}.$$

We are going to use Theorem 5.17 for Laplace exponents that are regularly varying at $\infty$ (resp. at 0). To be more specific we will assume that either (i)

$$\phi(\lambda) \sim \lambda^{\alpha/2}\ell(\lambda), \quad \lambda \to \infty, \tag{5.26}$$

where $0 < \alpha \le 2$, and $\ell$ is slowly varying at $\infty$, or (ii)

$$\phi(\lambda) \sim \lambda^{\alpha/2}\ell(\lambda), \quad \lambda \to 0, \tag{5.27}$$

where $0 < \alpha \le 2$, and $\ell$ is slowly varying at 0. In case (i), (5.26) implies $b > 0$ or $\mu(0,\infty) = \infty$. If $\phi$ is a special Bernstein function, then the corresponding subordinator $S$ has a decreasing potential density $u$ whose asymptotic behavior at 0 is then given by

$$u(t) \sim \frac{1}{\Gamma(\alpha/2)} \frac{t^{\alpha/2-1}}{\ell(1/t)}, \quad t \to 0+. \tag{5.28}$$

In case (ii), if $\phi$ is a special Bernstein function with $\lim_{\lambda \to \infty} \phi(\lambda) = \infty$, then the corresponding subordinator $S$ has a decreasing potential density $u$ whose asymptotic behavior at $\infty$ is then given by

$$u(t) \sim \frac{1}{\Gamma(\alpha/2)} \frac{t^{\alpha/2-1}}{\ell(1/t)}, \quad t \to \infty. \tag{5.29}$$

As consequences of the above, we immediately get the following: (1) for $\alpha \in (0, 2)$, the potential density of the relativistic $\alpha/2$-stable subordinator satisfies

$$u(t) \sim \frac{t^{\alpha/2-1}}{\Gamma(\alpha/2)}, \quad t \to 0+, \tag{5.30}$$

$$u(t) \sim \frac{\alpha}{2} m^{1-2/\alpha}, \quad t \to \infty; \tag{5.31}$$

(2) for $\alpha \in (0, 2)$, $\beta \in (0, 2 - \alpha)$, the potential density of the subordinator corresponding to Example 5.15 satisfies

$$u(t) \sim \frac{1}{\Gamma(\alpha/2)} \frac{1}{t^{1-\alpha/2}|\log t|^{\beta/2}}, \quad t \to 0+, \tag{5.32}$$

$$u(t) \sim \frac{1}{\Gamma(\alpha/2 + \beta/2)} \frac{1}{t^{1-(\alpha+\beta)/2}}, \quad t \to \infty; \tag{5.33}$$

(3) for $\alpha \in (0,2)$, $\beta \in (0,\alpha)$, the potential density of the subordinator corresponding to Example 5.16 satisfies

$$u(t) \sim \frac{\alpha}{2\Gamma(1+\alpha/2)} \frac{|\log t|^{\beta/2}}{t^{1-\alpha/2}}, \quad t \to 0+, \tag{5.34}$$

$$u(t) \sim \frac{\alpha-\beta}{2\Gamma(1+(\alpha-\beta)/2)} \frac{1}{t^{1-(\alpha-\beta)/2}}, \quad t \to \infty. \tag{5.35}$$

In the case when the subordinator has a positive drift $b > 0$, the potential density $u$ always exists, it is continuous, and $u(0+) = b$. For example, this will be the case when $\phi(\lambda) = b\lambda + \lambda^{\alpha/2}$. Recall (see Example 5.13) that the potential density is given by the rather explicit formula $u(t) = (1/b)E_{1-\alpha/2}(b^{2/(2-\alpha)}t)$ . The asymptotic behavior of $u(t)$ as $t \to \infty$ is not easily derived from this formula. On the other hand, since $\phi(\lambda) \sim \lambda^{\alpha/2}$ as $\lambda \to 0$, it follows from (5.29) that $u(t) \sim t^{\alpha/2-1}/\Gamma(\alpha/2)$ as $t \to \infty$.

Note that the gamma subordinator, geometric $\alpha/2$-stable subordinators, iterated geometric stable subordinators and Bessel subordinators have Laplace exponents that are not regularly varying with strictly positive exponent at $\infty$, but are rather slowly varying at $\infty$. In this case, Karamata's monotone density theorem cannot be used, and we need more refined versions of both Tauberian and monotone density theorems. The results are also taken from [21].

**Theorem 5.18.** *(a) (de Haan's Tauberian Theorem) Let $U : (0,\infty) \to (0,\infty)$ be an increasing function. If $\ell$ is slowly varying at $\infty$ (resp. at $0+$), $c \geq 0$, the following are equivalent:*

*(i) As $t \to \infty$ (resp. $t \to 0+$)*

$$\frac{U(\lambda t) - U(t)}{\ell(t)} \to c \log \lambda, \quad \forall \lambda > 0.$$

*(ii) As $t \to \infty$ (resp. $t \to 0+$)*

$$\frac{\mathcal{L}U(\frac{1}{\lambda t}) - \mathcal{L}U(\frac{1}{t})}{\ell(t)} \to c \log \lambda, \quad \forall \lambda > 0.$$

*(b) (de Haan's Monotone Density Theorem) If additionally $U(dx) = u(x)\,dx$, where $u$ is monotone and nonnegative, and $c > 0$, then (i) and (ii) are equivalent to:*

*(iii) As $t \to \infty$ (resp. $t \to 0+$)*

$$u(t) \sim ct^{-1}\ell(t).$$

We are going to apply this result to establish the asymptotic behaviors of the potential density of geometric stable subordinators, iterated geometric stable subordinators and Bessel subordinators at zero.

**Proposition 5.19.** *For any* $\alpha \in (0, 2]$, *let* $\phi(\lambda) = \log(1 + \lambda^{\alpha/2})$, *and let* $u$ *be the potential density of the corresponding subordinator. Then*

$$u(t) \sim \frac{2}{\alpha t (\log t)^2}, \quad t \to 0+,$$

$$u(t) \sim \frac{t^{\alpha/2-1}}{\Gamma(\alpha/2)}, \quad t \to \infty.$$

**Proof.** Recall that

$$\mathcal{L}U(\lambda) = 1/\phi(\lambda) = 1/\log(1 + \lambda^{\alpha/2}).$$

Since

$$\frac{\mathcal{L}U(\frac{1}{x\lambda}) - \mathcal{L}U(\frac{1}{\lambda})}{(\log \lambda)^{-2}} \to \frac{2}{\alpha} \log x, \quad \forall x > 0,$$

as $\lambda \to 0+$, we have by (the 0+ version of) Theorem 5.18 (a) that

$$\frac{U(xt) - U(t)}{(\log t)^{-2}} \to \frac{2}{\alpha} \log x, \quad x > 0,$$

as $t \to 0+$. Now we can apply (the 0+ version of) Theorem 5.18 (b) to get that

$$u(t) \sim \frac{2}{\alpha t (\log t)^2}$$

as $t \to 0+$. The asymptotic behavior of $u(t)$ as $t \to \infty$ follows from Theorem 5.17.                                                                                 □

In order to deal with the iterated geometric stable subordinators, let $e_0 = 0$, and inductively, $e_n = e^{e_{n-1}}$, $n \geq 1$. For $n \geq 1$ define $l_n : (e_n, \infty) \to (0, \infty)$ by

$$l_n(y) = \log \log \ldots \log y, \quad n \text{ times}. \tag{5.36}$$

Further, let $L_0(y) = 1$, and for $n \in \mathbb{N}$, define $L_n : (e_n, \infty) \to (0, \infty)$ by

$$L_n(y) = l_1(y) l_2(y) \ldots l_n(y). \tag{5.37}$$

Note that $l'_n(y) = 1/(y L_{n-1}(y))$ for every $n \geq 1$. Let $\alpha \in (0, 2]$ and recall from Example 5.12 that $\phi^{(1)}(y) := \log(1 + y^{\alpha/2})$, and for $n \geq 1$, $\phi^{(n)}(y) := \phi(\phi^{(n-1)}(y))$. Let $k_n(y) := 1/\phi^{(n)}(y)$.

**Lemma 5.20.** *Let* $t > 0$. *For every* $n \in \mathbb{N}$,

$$\lim_{y \to \infty} (k_n(ty) - k_n(y)) L_{n-1}(y) l_n(y)^2 = -\frac{2}{\alpha} \log t.$$

**Proof.** The proof for $n = 1$ is straightforward and is implicit in the proof of Proposition 5.19. We only give the proof for $n = 2$, the proof for general $n$ is similar. Using the fact that

$$\log(1 + y) \sim y, \quad y \to 0+, \tag{5.38}$$

we can easily get that

$$\lim_{y \to \infty} \left( \log \frac{\log y}{\log(yt)} \right) \log y = - \lim_{y \to \infty} \left( \log \frac{\log y + \log t}{\log y} \right) \log y = - \log t. \tag{5.39}$$

Using (5.38) and the elementary fact that $\log(1 + y) \sim \log y$ as $y \to \infty$ we get that

$$\lim_{y \to \infty} (k_2(ty) - k_2(y)) L_1(y) l_2(y)^2$$

$$= \frac{\alpha}{2} \lim_{y \to \infty} \left( \log \frac{\log(1 + y^{\alpha/2})}{\log(1 + (ty)^{\alpha/2})} \right)$$

$$\times \frac{\log y (\log \log y)^2}{(\alpha/2)^2 \log(\log(1 + y^{\alpha/2})) \log(\log(1 + (ty)^{\alpha/2}))}$$

$$= \frac{2}{\alpha} \lim_{y \to \infty} \left( \log \frac{\log y}{\log(yt)} \right) \log y = -\frac{2}{\alpha} \log t.$$

$\square$

Recall that $U^{(n)}$ denotes the potential measure and $u^{(n)}(t)$ the potential density of the iterated geometric stable subordinator $S^{(n)}$ with the Laplace exponent $\phi^{(n)}$.

**Proposition 5.21.** *For any $\alpha \in (0, 2]$, we have*

$$u^{(n)}(t) \sim \frac{2}{\alpha t L_{n-1}(\frac{1}{t}) l_n(\frac{1}{t})^2}, \quad t \to 0+, \tag{5.40}$$

$$u^{(n)}(t) \sim \frac{t^{(\alpha/2)^n - 1}}{\Gamma((\alpha/2)^n)}, \quad t \to \infty. \tag{5.41}$$

**Proof.** Using Lemma 5.20 we can easily see that

$$\frac{\mathcal{L}U^{(n)}(\frac{1}{x\lambda}) - \mathcal{L}U^{(n)}(\frac{1}{\lambda})}{(L_{n-1}(\frac{1}{\lambda}) l_n(\frac{1}{\lambda})^2)^{-1}} \to \frac{2}{\alpha} \log x, \quad \forall x > 0,$$

as $\lambda \to 0+$. Therefore, by (the 0+ version of) Theorem 5.18 (a) we have that

$$\frac{U^{(n)}(xt) - U^{(n)}(t)}{(L_{n-1}(\frac{1}{t}) l_n(\frac{1}{t})^2)^{-1}} \to \frac{2}{\alpha} \log x, \quad x > 0,$$

as $t \to 0+$. Now we can apply (the 0+ version of) Theorem 5.18 (b) to get that

$$u^{(n)}(t) \sim \frac{2}{\alpha t L_{n-1}(\frac{1}{t}) l_n(\frac{1}{t})^2}$$

as $t \to 0+$. The asymptotic behavior of $u^{(n)}(t)$ at $\infty$ follows easily from Theorem 5.17. □

Let $u_I$ and $u_K$ be the potential densities of the Bessel subordinators $I$ and $K$ respectively. Then we have the following result.

**Proposition 5.22.** *The potential densities of the Bessel subordinators satisfy the following asymptotics*

$$u_I(t) \sim \frac{1}{t(\log t)^2}, \quad t \to 0+,$$

$$u_K(t) \sim \frac{1}{t|\log t|^3}, \quad t \to 0+,$$

$$u_I(t) \sim \frac{1}{\sqrt{2\pi}} t^{-1/2}, \quad t \to \infty,$$

$$u_K(t) \sim 1, \quad t \to \infty.$$

**Proof.** The proofs of first two relations are direct applications of de Haan's Tauberian and monotone density theorems and the proofs of the last two are direct applications of Karamata's Tauberian and monotone density theorems. We omit the details. □

We now discuss the asymptotic behavior of the Lévy density of a subordinator.

**Proposition 5.23.** *Assume that the Laplace exponent $\phi$ of the subordinator $S$ is a complete Bernstein function and let $\mu(t)$ denote the density of its Lévy measure.*

*(i)   Let $0 < \alpha < 2$. If $\phi(\lambda) \sim \lambda^{\alpha/2} \ell(\lambda)$, $\lambda \to \infty$, and $\ell$ is a slowly varying function at $\infty$, then*

$$\mu(t) \sim \frac{\alpha/2}{\Gamma(1-\alpha/2)} t^{-1-\alpha/2} \ell(1/t), \quad t \to 0+ . \tag{5.42}$$

*(ii)   Let $0 < \alpha \leq 2$. If $\phi(\lambda) \sim \lambda^{\alpha/2} \ell(\lambda)$, $\lambda \to 0$, and $\ell$ is a slowly varying function at 0, then*

$$\mu(t) \sim \frac{\alpha/2}{\Gamma(1-\alpha/2)} t^{-1-\alpha/2} \ell(1/t), \quad t \to \infty. \tag{5.43}$$

**Proof.** (i) The assumption implies that there is no drift, $b = 0$, and hence by integration by parts,

$$\phi(\lambda) = \lambda \int_0^\infty e^{-\lambda t} \mu(t, \infty) \, dt \, .$$

Thus, $\int_0^\infty e^{-\lambda t} \mu(t, \infty) \, dt \sim \lambda^{\alpha/2-1} \ell(\lambda)$ as $\lambda \to \infty$, and (5.42) follows by first using Karamata's Tauberian theorem and then Karamata's monotone density theorem.

(ii) In this case it is possible that the drift $b$ is strictly positive, and thus

$$\phi(\lambda) = \lambda \left( b + \int_0^\infty e^{-\lambda t} \mu(t, \infty) \, dt \right) \, .$$

This implies that $\int_0^\infty e^{-\lambda t} \mu(t) \, dt \sim \lambda^{\alpha/2-1} \ell(\lambda)$ as $\lambda \to 0$, and (5.43) holds by Theorem 5.17.                                                                           $\square$

Note that if $\phi(\lambda) \sim b\lambda$ as $\lambda \to \infty$ and $b > 0$, nothing can be inferred about the behavior of the density $\mu(t)$ near zero. Next we record the asymptotic behavior of the Lévy density of the geometric stable subordinator. The first claim follows from (5.24), and the second from the previous proposition.

**Proposition 5.24.** *Let $\mu(dt) = \mu(t) \, dt$ be the Lévy measure of a geometrically $\alpha/2$-stable subordinator. Then*

(i)   *For $0 < \alpha \le 2$, $\mu(t) \sim \frac{\alpha}{2t}$, $t \to 0+$.*

(ii)  *For $0 < \alpha < 2$, $\mu(t) \sim \frac{\alpha/2}{\Gamma(1-\alpha/2)} t^{-\alpha/2-1}$, $t \to \infty$. For $\alpha = 2$, $\mu(t) = \frac{e^{-t}}{t}$.*

In the case of iterated geometric stable subordinators, we have only partial result for the asymptotic behavior of the density $\mu^{(n)}$ which follows from Proposition 5.43 (ii).

**Proposition 5.25.** *For any $\alpha \in (0, 2)$,*

$$\mu^{(n)}(t) \sim \frac{(\alpha/2)^n}{\Gamma(1 - (\alpha/2)^n)} t^{-1-(\alpha/2)^n}, \ t \to \infty \, .$$

**Remark 5.26.** *Note that we do not give the asymptotic behavior of $\mu^{(n)}(t)$ as $t \to \infty$ for $\alpha = 2$ (iterated gamma subordinator), and the asymptotic behavior of $\mu^{(n)}(t)$ as $t \to 0+$ for all $\alpha \in (0, 2]$. It is an open problem to determine the correct asymptotic behavior.*

The following results are immediate consequences of Proposition 5.23.

**Proposition 5.27.** *Suppose that $\alpha \in (0, 2)$ and $\beta \in (0, 2 - \alpha)$. Let $\mu(t)$ be the Lévy density of the subordinator corresponding to Example 5.15. Then*

$$\mu(t) \sim \frac{\alpha}{2\Gamma(1 - \alpha/2)} t^{-1-\alpha/2} (\log(1/t))^{\beta/2}, \quad t \to 0+,$$

$$\mu(t) \sim \frac{\alpha + \beta}{2\Gamma(1 - (\alpha + \beta)/2)} t^{-1-(\alpha+\beta)/2}, \quad t \to \infty.$$

**Proposition 5.28.** *Suppose that* $\alpha \in (0,2)$ *and* $\beta \in (0,\alpha)$. *Let* $\mu(t)$ *be the Lévy density of the subordinator corresponding to Example 5.16. Then*

$$\mu(t) \sim \frac{\alpha}{2\Gamma(1 - \alpha/2)} \frac{1}{t^{1+\alpha/2}(\log(1/t))^{\beta/2}}, \quad t \to 0+,$$

$$\mu(t) \sim \frac{\alpha - \beta}{2\Gamma(1 - (\alpha - \beta)/2)} \frac{1}{t^{1+(\alpha-\beta)/2}}, \quad t \to \infty.$$

We conclude this section with a discussion of the asymptotic behavior of transition densities of geometric stable subordinators. Let $S = (S_t : t \geq 0)$ be a geometric $\alpha/2$-stable subordinator, and let $(\eta_s : s \geq 0)$ be the corresponding convolution semigroup. Further, let $(\rho_s : s \geq 0)$ be the convolution semigroup corresponding to an $\alpha/2$-stable subordinator, and by abuse of notation, let $\rho_s$ denote the corresponding density. Then by (5.21) and the explicit formula (5.20), we see that $\eta_t$ has a density

$$f_s(t) = \int_0^\infty \rho_u(t) \frac{1}{\Gamma(s)} u^{s-1} e^{-u} \, du.$$

For $s = 1$, this formula reads

$$f_1(t) = \int_0^\infty \rho_u(t) e^{-u} \, du.$$

Moreover, we have shown in Example 5.11 that $f_1(t)$ is completely monotone. To be more precise, $f_1(t)$ is the density of the distribution function $F(t) = 1 - E_{\alpha/2}(t)$ of the probability measure $\eta_1$ (see (5.25)).

**Proposition 5.29.** *For any* $\alpha \in (0,2)$,

$$f_1(t) \sim \frac{1}{\Gamma(\alpha/2)} t^{\alpha/2-1}, \quad t \to 0+, \tag{5.44}$$

$$f_1(t) \sim 2\pi\Gamma\left(1 + \frac{\alpha}{2}\right) \sin\left(\frac{\alpha\pi}{4}\right) t^{-1-\frac{\alpha}{2}}, \quad t \to \infty. \tag{5.45}$$

**Proof.** The first relation follows from the explicit form of the distribution function $F(t) = 1 - E_{\alpha/2}(t)$ and Karamata's monotone density theorem. For the second relation, use the scaling property of stable distribution, $\rho_u(t) = u^{-2/\alpha}\rho_1(u^{-2/\alpha}t)$, to get

$$f_1(t) = \int_0^\infty e^{-u} u^{-2/\alpha} \rho_1(u^{-2/\alpha}t) \, du.$$

Now use (5.18), (5.19) and dominated convergence theorem to obtain the
required asymptotic behavior. □

## 5.3 Subordinate Brownian Motion

### 5.3.1 Definitions and Technical Lemma

Let $X = (X_t, \mathbb{P}^x)$ be a $d$-dimensional Brownian motion. The transition
densities $p(t, x, y) = p(t, y - x)$, $x, y \in \mathbb{R}^d, t > 0$, of $X$ are given by

$$p(t, x) = (4\pi t)^{-d/2} \exp\left(-\frac{|x|^2}{4t}\right).$$

The semigroup $(P_t : t \geq 0)$ of $X$ is defined by $P_t f(x) = \mathbb{E}^x[f(X_t)] = \int_{\mathbb{R}^d} p(t, x, y) f(y)\, dy$, where $f$ is a nonnegative Borel function on $\mathbb{R}^d$. Recall
that if $d \geq 3$, the Green function $G^{(2)}(x, y) = G^{(2)}(x - y)$, $x, y \in \mathbb{R}^d$, of $X$ is
well defined and is equal to

$$G^{(2)}(x) = \int_0^\infty p(t, x)\, dt = \frac{\Gamma(d/2 - 1)}{4\pi^{d/2}} |x|^{-d+2}.$$

Let $S = (S_t : t \geq 0)$ be a subordinator independent of $X$, with Laplace
exponent $\phi(\lambda)$, Lévy measure $\mu$, drift $b \geq 0$, no killing, potential measure $U$, and convolution semigroup $(\eta_t : t \geq 0)$. We define a new process
$Y = (Y_t : t \geq 0)$ by $Y_t := X(S_t)$. Then $Y$ is a Lévy process with characteris-
tic exponent $\Phi(x) = \phi(|x|^2)$ (see e.g. [135], pp.197–198) called a *subordinate
Brownian motion*. The semigroup $(Q_t : t \geq 0)$ of the process $Y$ is given by

$$Q_t f(x) = \mathbb{E}^x[f(Y_t)] = \mathbb{E}^x[f(X(S_t))] = \int_0^\infty P_s f(x)\, \eta_t(ds).$$

If the subordinator $S$ is not a compound Poisson process, then $Q_t$ has a
density $q(t, x, y) = q(t, x - y)$ given by $q(t, x) = \int_0^\infty p(s, x)\, \eta_t(ds)$.

From now on we assume that the subordinate process $Y$ is transient.
According to the criterion due to Port and Stone ([130]), $Y$ is transient if
and only if for some small $r > 0$, $\int_{|x|<r} \Re(\frac{1}{\Phi(x)})\, dx < \infty$. Since $\Phi(x) = \phi(|x|^2)$
is real, it follows that $Y$ is transient if and only if

$$\int_{0+} \frac{\lambda^{d/2-1}}{\phi(\lambda)}\, d\lambda < \infty. \tag{5.46}$$

This is always true if $d \geq 3$, and, depending on the subordinator, may be true for $d = 1$ or $d = 2$. For $x \in \mathbb{R}^d$ and $A$ Borel subset of $\mathbb{R}^d$, the occupation measure is given by

$$G(x, A) = \mathbb{E}^x \int_0^\infty 1_{(Y_t \in A)} = \int_0^\infty Q_t 1_A(x)\, dt = \int_0^\infty \int_0^\infty P_s 1_A(x) \eta_t(ds)\, dt$$

$$= \int_0^\infty P_s 1_A\, U(ds) = \int_A \int_0^\infty p(s, x, y)\, U(ds)\, dy\,,$$

where the second line follows from (5.6). If $A$ is bounded, then by the transience of $Y$, $G(x, A) < \infty$ for every $x \in \mathbb{R}^d$. Let $G(x, y)$ denote the density of the occupation measure $G(x, \cdot)$. Clearly, $G(x, y) = G(y - x)$ where

$$G(x) = \int_0^\infty p(t, x)\, U(dt) = \int_0^\infty p(t, x) u(t)\, dt\,, \tag{5.47}$$

and the last equality holds in case when $U$ has a potential density $u$.

The Lévy measure $\pi$ of $Y$ is given by (see e.g. [135], pp. 197–198)

$$\pi(A) = \int_A \int_0^\infty p(t, x)\, \mu(dt)\, dx = \int_A J(x)\, dx\,, \quad A \subset \mathbb{R}^d\,,$$

where

$$J(x) := \int_0^\infty p(t, x)\, \mu(dt) = \int_0^\infty p(t, x)\mu(t)dt\,, \tag{5.48}$$

is called the jumping function of $Y$. The last equality is valid in the case when $\mu(dt)$ has a density $\mu(t)$. Define the function $j : (0, \infty) \to (0, \infty)$ by

$$j(r) := \int_0^\infty (4\pi)^{-d/2} t^{-d/2} \exp\left(-\frac{r^2}{4t}\right) \mu(dt)\,, \quad r > 0\,, \tag{5.49}$$

and note that by (5.48), $J(x) = j(|x|)$, $x \in \mathbb{R}^d \setminus \{0\}$. We state the following well-known conditions describing when a Lévy process is a subordinate Brownian motion (for a proof, see e.g. [97], pp. 190–192).

**Proposition 5.30.** *Let $Y = (Y_t : t \geq 0)$ be a $d$-dimensional Lévy process with the characteristic triple $(b, A, \pi)$. Then $Y$ is a subordinate Brownian motion if and only if $\pi$ has a rotationally invariant density $x \mapsto j(|x|)$ such that $r \mapsto j(\sqrt{r})$ is a completely monotone function on $(0, \infty)$, $A = cI_d$ with $c \geq 0$, and $b = 0$.*

**Example 5.31.** (i) Let $\phi(\lambda) = \lambda^{\alpha/2}$, $0 < \alpha < 2$, and let $S$ be the corresponding $\alpha/2$-stable subordinator. The characteristic exponent of the subordinate process $Y$ is equal to $\Phi(x) = \phi(|x|^2) = |x|^\alpha$. Hence $Y$ is a rotationally invariant $\alpha$-stable process. From now on we will (imprecisely) refer to this process as a symmetric $\alpha$-stable process. $Y$ is transient if and only if $d > \alpha$.

The jumping function of $Y$ is given by

$$J(x) = \frac{\alpha 2^{\alpha-1}\Gamma(\frac{d+\alpha}{2})}{\pi^{d/2}\Gamma(1-\frac{\alpha}{2})} |x|^{-\alpha-d}, \quad x \in \mathbb{R}^d,$$

and when $d \geq 3$, the Green function of $Y$ is given by the Riesz kernel

$$G(x) = \frac{1}{\pi^{d/2}2^{\alpha}} \frac{\Gamma(\frac{d-\alpha}{2})}{\Gamma(\frac{\alpha}{2})} |x|^{\alpha-d}, \quad x \in \mathbb{R}^d.$$

(ii) For $0 < \alpha < 2$ and $m > 0$, let $\phi(\lambda) = (\lambda + m^{\alpha/2})^{2/\alpha} - m$, and let $S$ be the corresponding relativistic $\alpha/2$-stable subordinator. The characteristic exponent of the subordinate process $Y$ is equal to $\Phi(x) = \phi(|x|^2) = (|x|^2 + m^{\alpha/2})^{2/\alpha} - m$. The process $Y$ is called the *symmetric relativistic $\alpha$-stable process*. $Y$ is transient if and only if $d > 2$.

(iii) Let $\phi(\lambda) = \log(1 + \lambda)$, and let $S$ be the corresponding gamma subordinator. The characteristic exponent of the subordinate process $Y$ is given by $\Phi(x) = \log(1 + |x|^2)$. The process $Y$ is known in some finance literature (see [121] and [77]) as a *variance gamma* process (at least for $d = 1$). $Y$ is transient if and only if $d > 2$.

(iv) For $0 < \alpha < 2$, let $\phi(\lambda) = \log(1 + \lambda^{\alpha/2})$, and let $S$ be the corresponding subordinator. The characteristic exponent of the subordinate process $Y$ is given by $\Phi(x) = \log(1 + |x|^{\alpha})$. The process $Y$ is known as a rotationally invariant *geometric $\alpha$-stable process*. From now on we will (imprecisely) refer to this process as a symmetric geometric $\alpha$-stable process. $Y$ is transient if and only if $d > \alpha$.

(v) For $0 < \alpha < 2$, let $\phi^{(1)}(\lambda) = \log(1 + \lambda^{\alpha/2})$, and for $n > 1$, let $\phi^{(n)}(\lambda) = \phi^{(1)}(\phi^{(n-1)}(\lambda))$. Let $S^{(n)}$ be the corresponding iterated geometric stable subordinator. Denote $Y_t^{(n)} = X(S_t^{(n)})$. $Y^{(n)}$ is transient if and only if $d > 2(\alpha/2)^n$.

(vi) For $0 < \alpha < 2$ let $\phi(\lambda) = b\lambda + \lambda^{\alpha/2}$, and let $S$ be the corresponding subordinator. The characteristic exponent of the subordinate process $Y$ is $\Phi(x) = b|x|^2 + |x|^{\alpha}$. Hence $Y$ is the sum of a (multiple of) Brownian motion and an independent $\alpha$-stable process. Similarly, we can realize the sum of an $\alpha$-stable and an independent $\beta$-stable processes by subordinating Brownian motion $X$ with a subordinator having the Laplace exponent $\phi(\lambda) = \lambda^{\alpha/2} + \lambda^{\beta/2}$.

(vii) The characteristic exponent of the subordinate Brownian motion with the Bessel subordinator $S_I$ is $\log((1 + |x|^2) + \sqrt{(1 + |x|^2)^2 - 1})$ and so this process is transient if and only if $d > 1$. The characteristic exponent of the subordinate Brownian motion with the Bessel subordinator $S_K$ is $\frac{1}{2}(\log((1 + |x|^2) + \sqrt{(1 + |x|^2)^2 - 1}))^2$ and so this process is transient if and only if $d > 2$.

(viii) For $\alpha \in (0, 2), \beta \in (0, 2 - \alpha)$, let $S$ be the subordinator with Laplace exponent $\phi(\lambda) = \lambda^{\alpha/2}(\log(1 + \lambda))^{\beta/2}$. The characteristic exponent of the subordinate process $Y$ is $\Phi(x) = |x|^{\alpha}(\log(1 + |x|^2))^{\beta/2}$. $Y$ is transient if and only if $d > \alpha + \beta$.

(ix) For $\alpha \in (0,2), \beta \in (0,\alpha)$, let $S$ be the subordinator with Laplace exponent $\phi(\lambda) = \lambda^{\alpha/2}(\log(1+\lambda))^{-\beta/2}$. The characteristic exponent of the subordinate process $Y$ is $\Phi(x) = |x|^\alpha(\log(1+|x|^2))^{-\beta/2}$. $Y$ is transient if and only if $d > \alpha - \beta$.

In order to establish the asymptotic behaviors of the Green function $G$ and the jumping function $J$ of the subordinate Brownian motion $Y$, we start by defining an auxiliary function. For any slowly varying function $\ell$ at infinity and any $\xi > 0$, let

$$f_{\ell,\xi}(y,t) := \begin{cases} \frac{\ell(1/y)}{\ell(4t/y)}, & y < \frac{t}{\xi}, \\ 0, & y \geq \frac{t}{\xi}. \end{cases}$$

Now we state and prove the key technical lemma.

**Lemma 5.32.** *Suppose that $w : (0,\infty) \to (0,\infty)$ is a decreasing function satisfying the following two assumptions:*

(i)   *There exist constants $c_0 > 0$ and $\beta \in [0,2]$ with $\beta > 1 - d/2$, and a continuous functions $\ell : (0,\infty) \to (0,\infty)$ slowly varying at $\infty$ such that*

$$w(t) \sim \frac{c_0}{t^\beta \ell(1/t)}, \quad t \to 0+ . \tag{5.50}$$

(ii)  *If $d = 1$ or $d = 2$, then there exist a constant $c_\infty > 0$ and a constant $\gamma < d/2$ such that*

$$w(t) \sim c_\infty t^{\gamma-1}, \quad t \to +\infty . \tag{5.51}$$

*Let $g : (0,\infty) \to (0,\infty)$ be a function such that*

$$\int_0^\infty t^{d/2-2+\beta} e^{-t} g(t)\, dt < \infty .$$

*If there is $\xi > 0$ such that $f_{\ell,\xi}(y,t) \leq g(t)$ for all $y, t > 0$, then*

$$I(x) := \int_0^\infty (4\pi t)^{-d/2} e^{-\frac{|x|^2}{4t}} w(t)\, dt$$

$$\sim \frac{c_0 \Gamma(d/2 + \beta - 1)}{4^{1-\beta} \pi^{d/2}} \frac{1}{|x|^{d+2\beta-2}\, \ell(\frac{1}{|x|^2})}, \quad |x| \to 0 .$$

**Proof.** Let us first note that the assumptions of the lemma guarantee that $I(x) < \infty$ for every $x \neq 0$. By a change of variable we get

$$\int_0^\infty (4\pi t)^{-d/2} e^{-\frac{|x|^2}{4t}} w(t)\, dt = \frac{|x|^{-d+2}}{4\pi^{d/2}} \int_0^\infty t^{d/2-2} e^{-t} w\left(\frac{|x|^2}{4t}\right) dt$$

$$= \frac{1}{4\pi^{d/2}} \left( |x|^{-d+2} \int_0^{\xi|x|^2} + |x|^{-d+2} \int_{\xi|x|^2}^\infty \right)$$

$$= \frac{1}{4\pi^{d/2}} \left( |x|^{-d+2} I_1 + |x|^{-d+2} I_2 \right) .$$

We first consider $I_1$ for the case $d = 1$ or $d = 2$. It follows from the assumptions that there exists a positive constant $c_1$ such that $w(s) \leq c_1 s^{\gamma-1}$ for all $s \geq 1/(4\xi)$. Thus

$$I_1 \leq \int_0^{\xi|x|^2} t^{d/2-2} e^{-t} c_1 \left( \frac{|x|^2}{4t} \right)^{\gamma-1} dt$$

$$\leq c_2 |x|^{2\gamma-2} \int_0^{\xi|x|^2} t^{d/2-\gamma-1} dt = c_3 |x|^{d-2}.$$

It follows that

$$\lim_{|x| \to 0} \frac{|x|^{-d+2} I_1}{\frac{1}{|x|^{d+2\beta-2} \ell(\frac{1}{|x|^2})}} = 0. \tag{5.52}$$

In the case $d \geq 3$, we proceed similarly, using the bound $w(s) \leq w(1/(4\xi))$ for $s \geq 1/(4\xi)$.

Now we consider $I_2$:

$$|x|^{-d+2} I_2 = \frac{1}{|x|^{d-2}} \int_{\xi|x|^2}^{\infty} t^{d/2-2} e^{-t} w \left( \frac{|x|^2}{4t} \right) dt$$

$$= \frac{4^\beta}{|x|^{d+2\beta-2} \ell(\frac{1}{|x|^2})} \int_{\xi|x|^2}^{\infty} t^{d/2-2+\beta} e^{-t} \frac{w \left( \frac{|x|^2}{4t} \right)}{\left( \frac{|x|^2}{4t} \right)^\beta \ell(\frac{4t}{|x|^2})} \frac{\ell(\frac{1}{|x|^2})}{\ell(\frac{4t}{|x|^2})} dt.$$

Using the assumption (5.50), we can see that there is a constant $c > 0$ such that

$$\frac{w \left( \frac{|x|^2}{4t} \right)}{\left( \frac{|x|^2}{4t} \right)^\beta \ell(\frac{4t}{|x|^2})} < c$$

for all $t$ and $x$ satisfying $|x|^2/(4t) \leq 1/(4\xi)$. Since $\ell$ is slowly varying at infinity,

$$\lim_{|x| \to 0} \frac{\ell(\frac{1}{|x|^2})}{\ell(\frac{4t}{|x|^2})} = 1$$

for all $t > 0$. Note that

$$\frac{\ell(\frac{1}{|x|^2})}{\ell(\frac{4t}{|x|^2})} = f_{\ell,\xi}(|x|^2, t).$$

It follows from the assumption that

$$t^{d/2-2+\beta} e^{-t} \frac{w \left( \frac{|x|^2}{4t} \right)}{\left( \frac{|x|^2}{4t} \right)^\beta \ell(\frac{4t}{|x|^2})} \frac{\ell(\frac{1}{|x|^2})}{\ell(\frac{4t}{|x|^2})} \leq c t^{d/2-2+\beta} e^{-t} g(t).$$

Therefore, by the dominated convergence theorem we have

$$\lim_{|x| \to 0} \int_{\xi |x|^2}^{\infty} t^{d/2-2+\beta} e^{-t} \frac{w\left(\frac{|x|^2}{4t}\right)}{\left(\frac{|x|^2}{4t}\right)^\beta \ell\left(\frac{4t}{|x|^2}\right)} \frac{\ell\left(\frac{1}{|x|^2}\right)}{\ell\left(\frac{4t}{|x|^2}\right)} \, dt = \int_0^\infty c_0 t^{d/2-2+\beta} e^{-t} \, dt$$

$$= c_0 \Gamma(d/2 + \beta - 1).$$

Hence,

$$\lim_{|x| \to 0} \frac{|x|^{-d+2} I_2}{\frac{4^\beta}{|x|^{d+2\beta-2} \ell\left(\frac{1}{|x|^2}\right)}} = c_0 \Gamma(d/2 + \beta - 1). \tag{5.53}$$

Finally, combining (5.52) and (5.53) we get

$$\lim_{|x| \to 0} \frac{I(x)}{\frac{1}{|x|^{d+2\beta-2} \ell\left(\frac{1}{|x|^2}\right)}} = \frac{c_0 \Gamma(d/2 + \beta - 1)}{4^{1-\beta} \pi^{d/2}}.$$

$\square$

**Remark 5.33.** *Note that if in (5.50) we have that $\ell = 1$, then $f_{\ell,\xi} = 1$, hence $f_{\ell,\xi}(y,t) \le g(y)$ and $\int_0^\infty t^{d/2+\beta-2} e^{-t} g(t) \, dt < \infty$ with $g = 1$ provided $\beta > 1 - d/2$.*

## 5.3.2  Asymptotic Behavior of the Green Function

The goal of this subsection is to establish the asymptotic behavior of the Green function $G(x)$ of the subordinate process $Y$ under certain assumptions on the Laplace exponent of the subordinator $S$. We start with the asymptotic behavior when $|x| \to 0$ for the following cases: (1) $\phi(\lambda)$ has a power law behavior at $\infty$, (2) $S$ is a geometric $\alpha/2$-stable subordinator, $0 < \alpha \le 2$, (3) $S$ is an iterated geometric stable subordinator, (4) $S$ is a Bessel subordinator, and (v) $S$ is the subordinator corresponding to Example 5.15 or Example 5.16.

**Theorem 5.34.** *Suppose that $S = (S_t : t \ge 0)$ is a subordinator whose Laplace exponent $\phi(\lambda) = b\lambda + \int_0^\infty (1 - e^{-\lambda t}) \, \mu(dt)$ satisfies one of the following two assumptions:*

*(i)   $b > 0$,*
*(ii)  $S$ is a special subordinator and $\phi(\lambda) \sim \gamma^{-1} \lambda^{\alpha/2}$ as $\lambda \to \infty$, for $0 < \alpha < 2$.*

*If $Y$ is transient, then*

$$G(x) \sim \frac{\gamma}{\pi^{d/2} 2^\alpha} \frac{\Gamma(\frac{d-\alpha}{2})}{\Gamma(\frac{\alpha}{2})} |x|^{\alpha-d}, \quad |x| \to 0, \tag{5.54}$$

*(where in case (i), $\gamma^{-1} = b$ and $\alpha = 2$).*

**Proof.** (i) In this case, $\phi(\lambda) \sim b\lambda$, $\lambda \to \infty$. By Proposition 5.7, the potential measure $U$ has a continuous density $u$ satisfying $u(0+) = 1/b = \gamma$ and $u(t) \le u(0+)$ for all $t > 0$. Note first that by change of variables

$$\int_0^\infty (4\pi t)^{-d/2} \exp\left(-\frac{|x|^2}{4t}\right) u(t)\, dt = \frac{|x|^{-d+2}}{4\pi^{d/2}} \int_0^\infty s^{d/2-2} e^{-s} u\left(\frac{|x|^2}{4s}\right) ds. \tag{5.55}$$

By Proposition 5.7, $\lim_{x\to 0} u(|x|^2/(4s)) = u(0+) = \gamma$ for all $s > 0$ and $u(|x|^2/(4s))$ is bounded by $u(0+)$. Hence, by the bounded convergence theorem,

$$\lim_{x\to 0} \frac{1}{|x|^{-d+2}} \int_0^\infty (4\pi t)^{-d/2} \exp\left(-\frac{|x|^2}{4t}\right) u(t)\, dt = \frac{\gamma \Gamma(d/2 - 1)}{4\pi^{d/2}}. \tag{5.56}$$

(ii) In this case the potential measure $U$ has a decreasing density $u$ which by (5.28) satisfies

$$u(t) \sim \frac{\gamma}{\Gamma(\alpha/2)} \frac{1}{t^{1-\alpha/2}}, \quad t \to 0+ .$$

By recalling Remark 5.33, we can now apply Lemma 5.32 with $\beta = 1 - \alpha/2$ to obtain the required asymptotic behavior. $\qquad\Box$

**Theorem 5.35.** *For any $\alpha \in (0, 2]$, let $\phi(\lambda) = \log(1 + \lambda^{\alpha/2})$ and let $S$ be the corresponding geometric $\alpha/2$-stable subordinator. If $d > \alpha$, then the Green function of the subordinate process $Y$ satisfies*

$$G(x) \sim \frac{\Gamma(d/2)}{2\alpha\pi^{d/2}|x|^d \log^2 \frac{1}{|x|}}, \quad |x| \to 0. \tag{5.57}$$

**Proof.** We apply Lemma 5.32 with $w(t) = u(t)$, the potential density of $S$. By Proposition 5.19, $u(t) \sim \frac{2}{\alpha t \log^2 t}$ as $t \to 0+$, so we take $c_0 = 2/\alpha$, $\beta = 1$ and $\ell(t) = \log^2 t$. Moreover, by the second part of Proposition 5.19, $u(t) \sim t^{\alpha/2-1}/(\Gamma(\alpha)/2)$ as $t \to +\infty$, so we can take $\gamma = \alpha/2 < d/2$. Choose $\xi = 1/2$. Let

$$f(y, t) := f_{\ell, 1/2}(y, t) = \begin{cases} \dfrac{\log^2 y}{\log^2 \frac{y}{4t}}, & y < 2t, \\ 0, & y \ge 2t. \end{cases}$$

Define

$$g(t) := \begin{cases} \dfrac{\log^2 2t}{\log^2 2}, & t < \frac{1}{4}, \\ 1, & t \ge \frac{1}{4}. \end{cases}$$

In order to show that $f(y, t) \le g(t)$, first let $t < 1/4$. Then $y \mapsto f(y, t)$ is an increasing function for $0 < y < 2t$. Hence,

$$\sup_{0 < y < 2t} f(y, t) = f(2t, t) = \frac{\log^2 2t}{\log^2 2}.$$

Clearly, $f(y, 1/4) = 1$. For $t > 1/4$, $y \mapsto f(y, t)$ is a decreasing function for $0 < y < 1$. Hence

$$\sup_{0 < y < (2t) \wedge 1} f(y, t) = f(0, t) := \lim_{y \to 0} f(y, t) = 1.$$

For $t > 1/2$, elementary consideration gives that

$$\sup_{1 < y < 2t} f(y, t) \le \frac{\log^2 2t}{\log^2 2}.$$

Clearly,

$$\int_0^\infty t^{d/2-1} e^{-t} g(t)\, dt < \infty,$$

and the required asymptotic behavior follows from Lemma 5.32.          □

For $n \ge 1$, let $S^{(n)}$ be the iterated geometric stable subordinator with the Laplace exponent $\phi^{(n)}$. Recall that $\phi^{(1)}(\lambda) = \log(1 + \lambda^{\alpha/2})$, $0 < \alpha \le 2$, and $\phi^{(n)} = \phi^{(1)} \circ \phi^{(n-1)}$. Let $Y_t^{(n)} = X(S_t^{(n)})$ be the subordinate process and assume that $d > 2(\alpha/2)^n$. Denote the Green function of $Y^{(n)}$ by $G^{(n)}$. We want to study the asymptotic behavior of $G^{(n)}$ using Lemma 5.32. In order to check the conditions of that lemma, we need some preparations.

For $n \in \mathbb{N}$, define $f_n : (0, 1/e_n) \times (0, \infty) \to [0, \infty)$ by

$$f_n(y, t) := \begin{cases} \dfrac{L_{n-1}(\frac{1}{y}) l_n(\frac{1}{y})^2}{L_{n-1}(\frac{4t}{y}) l_n(\frac{4t}{y})^2}, & y < \frac{2t}{e_n}, \\[2mm] 0, & y \ge \frac{2t}{e_n}. \end{cases}$$

Note that $f_n$ is equal to the function $f_{\ell,\xi}$, defined before Lemma 5.32, with $\ell(y) = L_{n-1}(y) l_n(y)^2$ and $\xi = e_n/2$. Also, for $n \in \mathbb{N}$, let

$$g_n(t) := \begin{cases} f_n(\frac{2t}{e_n}, t), & t < 1/4, \\ 1, & t \ge 1/4. \end{cases}$$

Moreover, for $n \in \mathbb{N}$, define $h_n : (0, 1/e_n) \times (0, \infty) \to (0, \infty)$ by

$$h_n(y, t) := \frac{l_n(\frac{1}{y})}{l_n(\frac{4t}{y})}.$$

Clearly, for $0 < y < \frac{2t}{e_n} \wedge \frac{1}{e_n}$ we have that

$$f_n(y, t) = h_1(y, t) \ldots h_{n-1}(y, t) h_n(y, t)^2. \tag{5.58}$$

**Lemma 5.36.** *For all $y \in (0, 1/e_n)$ and all $t > 0$ we have $f_n(y, t) \leq g_n(t)$. Moreover, $\int_0^\infty t^{d/2-1} e^{-t} g_n(t) \, dt < \infty$.*

**Proof.** A direct calculation of partial derivative gives

$$\frac{\partial h_n}{\partial y}(y, t) = \frac{L_n(\frac{1}{y}) - L_n(\frac{4t}{y})}{y L_{n-1}(\frac{1}{y}) L_{n-1}(\frac{4t}{y}) l_n(\frac{4t}{y})^2}.$$

The denominator is always positive. Clearly, the numerator is positive if and only if $t > 1/4$. Therefore, for $t < 1/4$, $y \mapsto h_n(y, t)$ is increasing on $(0, 2t/e_n)$, while for $t > 1/4$ it is decreasing on $(0, 2t/e_n)$.

Let $t < 1/4$. It follows from (5.58) and the fact that $y \mapsto h_n(y, t)$ is increasing on $(0, 2t/e_n)$ that $y \mapsto f_n(y, t)$ is increasing for $0 < y < 2t/e_n$. Therefore,

$$\sup_{0 < y < 2t/e_n} f_n(y, t) \leq f_n(2t/e_n, t) = g_n(t).$$

Clearly, $f_n(y, 1/4) = 1$. For $y \geq 1/4$, it follows from (5.58) and the fact that $y \mapsto h_n(y, t)$ is decreasing on $(0, 2t/e_n)$ that $y \mapsto f_n(y, t)$ is decreasing for $0 < y < 1/e_n$. Hence

$$\sup_{0 < y < \frac{2t}{e_n} \wedge \frac{1}{e_n}} f_n(y, t) = f(0, t) := \lim_{y \to 0} f_n(y, t) = 1.$$

For $t > 1/2$, elementary consideration gives that

$$\sup_{\frac{1}{e_n} < y < \frac{2t}{e_n} \wedge \frac{1}{e_n}} f_n(y, t) \leq g_n(t).$$

The integrability statement of the lemma is obvious.                    $\square$

**Theorem 5.37.** *If $d > 2(\alpha/2)^n$, we have*

$$G^{(n)}(x) \sim \frac{\Gamma(d/2)}{2\alpha \pi^{d/2} |x|^d L_{n-1}(1/|x|^2) l_n(1/|x|^2)^2}, \quad |x| \to 0.$$

**Proof.** We apply Lemma 5.32 with $w(t) = u^{(n)}(t)$, the potential density of $S^{(n)}$. By Proposition 5.21,

$$u^{(n)}(t) \sim \frac{2}{\alpha t L_{n-1}(1/t) l_n(1/t)^2}, \quad t \to 0+,$$

so we take $c_0 = 2/\alpha$, $\beta = 1$ and $\ell(t) = L_{n-1}(t) l_n(t)^2$. By the second part of Proposition 5.21, $u^{(n)}(t)$ is of order $t^{(\alpha/2)^n - 1}$ as $t \to \infty$, so we may take $\gamma = (\alpha/2)^n < d/2$. Choose $\xi = e_n/2$. The result follows from Lemma 5.32 and Lemma 5.36                                                          $\square$

Using arguments similar to that used in the proof of Theorem 5.35, together with Proposition 5.22, (5.32) and (5.34), we can easily get the following two results.

**Theorem 5.38.** *(i) Suppose $d > 1$. Let $G_I$ be the Green function of the subordinate Brownian motion via the Bessel subordinator $S_I$. Then*

$$G_I(x) \sim \frac{\Gamma(d/2)}{4\pi^{d/2}|x|^d \log^2 \frac{1}{|x|}}, \quad |x| \to 0.$$

*(ii) Suppose $d > 2$. Let $G_K$ be the Green function of the subordinate Brownian motion via the Bessel subordinator $S_K$. Then*

$$G_K(x) \sim \frac{\Gamma(d/2)}{4\pi^{d/2}|x|^d \log^3 \frac{1}{|x|}}, \quad |x| \to 0.$$

**Theorem 5.39.** *Suppose $\alpha \in (0,2), \beta \in (0, 2 - \alpha)$ and that $S$ is the subordinator corresponding Example 5.15. If $d > \alpha + \beta$, the Green function of the subordinate Brownian motion via $S$ satisfies*

$$G(x) \sim \frac{\alpha\Gamma((d-\alpha)/2}{2^{\alpha+1}\pi^{d/2}\Gamma(1+\alpha/2)} \frac{1}{|x|^{d-\alpha}(\log(1/|x|^2))^{\beta/2}}, \quad |x| \to 0.$$

**Theorem 5.40.** *Suppose $\alpha \in (0,2)$, $\beta \in (0, \alpha)$ and that $S$ is the subordinator corresponding Example 5.16. If $d > \alpha - \beta$, the Green function of the subordinate Brownian motion via $S$ satisfies*

$$G(x) \sim \frac{\alpha\Gamma((d-\alpha)/2}{2^{\alpha+1}\pi^{d/2}\Gamma(1+\alpha/2)} \frac{(\log(1/|x|^2))^{\beta/2}}{|x|^{d-\alpha}}, \quad |x| \to 0.$$

**Proof.** The proof of this theorem is similar to that of Theorem 5.35, the only difference is that in this case when applying Lemma 5.32 we take the slowly varying function $\ell$ to be

$$\ell(t) = \begin{cases} (\log^2 t)^{-\beta/4}, \ t \geq 2, \\ (\log^2 2)^{-\beta/4}, \ t \leq 2. \end{cases}$$

Then using argument similar to that in the proof of Theorem 5.35 we can show that with the functions defined by

$$f(y,t) = \begin{cases} \frac{\ell(1/y)}{\ell(4t/y)}, \ y < 2t, \\ 0, \qquad y \geq 2t \end{cases} = \begin{cases} \left(\frac{\log^2(4t/y)}{\log^2(1/y)}\right)^{\beta/4}, \ t < 1/4, \ y < 2t, \\ \left(\frac{\log^2(4t/y)}{\log^2(1/y)}\right)^{\beta/4}, \ t \geq 1/4, \ y < 1/2, \\ \left(\frac{\log^2(4t/y)}{\log^2 2}\right)^{\beta/4}, \ t < 1/4, \ 1/2 < y < 2t, \\ 0, \qquad\qquad y \geq 2t, \end{cases}$$

and

$$g(t) = \begin{cases} \left(\frac{\log^2(8t)}{\log^2 2}\right)^{\beta/4}, & t > 1/4, \\ 1, & t \le 1/4. \end{cases}$$

we have $f(y,t) \le g(t)$ for all $y > 0$ and $t > 0$. The rest of the proof is exactly the same as that of Theorem 5.35. $\qquad\square$

By using results and methods developed so far, we can obtain the following table of the asymptotic behavior of the Green function of the subordinate Brownian motion depending on the Laplace exponent of the subordinator. The left column contains Laplace exponents, while the right column describes the asymptotic behavior of $G(x)$ as $|x| \to 0$, up to a constant.

| Laplace exponent $\phi$ | Green function $G \sim c \cdot$ |
|---|---|
| $\lambda$ | $|x|^{-d} \, |x|^2$ |
| $\int_0^1 \lambda^{1-\beta}\beta^\eta d\beta \quad (\eta > -1)$ | $|x|^{-d} \, |x|^2 \log(\frac{1}{|x|^2})^{\eta+1}$ |
| $\lambda^{\alpha/2}(\log(1+\lambda))^{\beta/2}, 0 < \alpha < 2, 0 < \beta < 2 - \alpha,$ | $|x|^{-d} \, |x|^\alpha \frac{1}{(\log(1/|x|^2))^{\beta/2}}$ |
| $\lambda^{\alpha/2}, 0 < \alpha < 2$ | $|x|^{-d} \, |x|^\alpha$ |
| $\lambda^{\alpha/2}(\log(1+\lambda))^{-\beta/2}, 0 < \alpha < 2, 0 < \beta < \alpha,$ | $|x|^{-d} \, |x|^\alpha (\log(1/|x|^2))^{\beta/2}$ |
| $\log(1+\lambda^{\alpha/2}), 0 < \alpha \le 2$ | $|x|^{-d} \, \frac{1}{\log^2 \frac{1}{|x|^2}}$ |
| $\phi^{(n)}(\lambda)$ | $|x|^{-d} \, \frac{1}{L_{n-1}(\frac{1}{x})l_n(\frac{1}{x})^2}$ |

Notice that the singularity of the Green function increases from top to bottom. This is, of course, a consequence of the fact that the corresponding subordinator becomes slower and slower, hence the subordinate process $Y$ moves also more slowly for small times.

We look now at the asymptotic behavior of the Green function $G(x)$ for $|x| \to \infty$.

**Theorem 5.41.** *Suppose that $S = (S_t : t \ge 0)$ is a subordinator whose Laplace exponent*

$$\phi(\lambda) = b\lambda + \int_0^\infty (1 - e^{-\lambda t}) \, \mu(dt)$$

*is a special Bernstein function such that $\lim_{\lambda \to \infty} \phi(\lambda) = \infty$. If $\phi(\lambda) \sim \gamma^{-1}\lambda^{\alpha/2}$ as $\lambda \to 0+$ for $\alpha \in (0,2]$ with $\alpha < d$ and a positive constant $\gamma$, then*

$$G(x) \sim \frac{\gamma}{\pi^{d/2}2^\alpha} \frac{\Gamma(\frac{d-\alpha}{2})}{\Gamma(\frac{\alpha}{2})} |x|^{\alpha-d}$$

*as $|x| \to \infty$.*

**Proof.** By Theorem 5.1 the potential measure of the subordinator has a decreasing density. By use of Theorem 5.17, the assumption $\phi(\lambda) \sim \gamma^{-1}\lambda^{\alpha/2}$ as $\lambda \to 0+$ implies that

$$u(t) \sim \frac{\gamma}{\Gamma(\alpha/2)} t^{\alpha/2-1}, \quad t \to \infty.$$

Since $u$ is decreasing and integrable near 0, it is easy to show that there exists $t_0 > 0$ such that $u(t) \leq t^{-1}$ for all $t \in (0, t_0)$. Hence, we can find a positive constant $C$ such that

$$u(t) \leq C(t^{-1} \vee t^{\alpha/2-1}). \tag{5.59}$$

By change of variables we have

$$\int_0^\infty (4\pi t)^{-d/2} \exp\left(-\frac{|x|^2}{4t}\right) u(t)\, dt$$

$$= \frac{1}{4\pi^{d/2}} |x|^{-d+2} \int_0^\infty s^{d/2-2} e^{-s} u\left(\frac{|x|^2}{4s}\right) ds$$

$$= \frac{\gamma}{4\pi^{d/2}\Gamma(\alpha/2)} |x|^{-d+\alpha} \int_0^\infty s^{d/2-2} e^{-s} \frac{u\left(\frac{|x|^2}{4s}\right)}{\frac{\gamma}{\Gamma(\alpha/2)}\left(\frac{|x|^2}{4s}\right)^{\alpha/2-1}} \left(\frac{1}{4s}\right)^{\alpha/2-1} ds$$

$$= \frac{\gamma}{2^\alpha \pi^{d/2}\Gamma(\alpha/2)} |x|^{-d+\alpha} \int_0^\infty s^{d/2-\alpha/2-1} e^{-s} \frac{u\left(\frac{|x|^2}{4s}\right)}{\frac{\gamma}{\Gamma(\alpha/2)}\left(\frac{|x|^2}{4s}\right)^{\alpha/2-1}} ds.$$

Let $|x| \geq 2$. Then by (5.59),

$$\frac{u\left(\frac{|x|^2}{4s}\right)}{\left(\frac{|x|^2}{4s}\right)^{\alpha/2-1}} \leq C\left(\left(\frac{|x|^2}{4s}\right)^{-\alpha/2} \vee 1\right) \leq C(s^{\alpha/2} \vee 1).$$

It follows that the integrand in the last formula above is bounded by an integrable function, so we may use the dominated convergence theorem to obtain

$$\lim_{|x|\to\infty} \frac{1}{|x|^{-d+\alpha}} \int_0^\infty (4\pi t)^{-d/2} \exp\left(-\frac{|x|^2}{4t}\right) u(t)\, dt = \frac{\gamma}{2^\alpha \pi^{d/2}} \frac{\Gamma(\frac{d-\alpha}{2})}{\Gamma(\frac{\alpha}{2})},$$

which proves the result.                                                        □

Examples of subordinators that satisfy the assumptions of the last theorem are relativistic $\beta/2$-stable subordinators (with $\alpha$ in the theorem equal to 2), gamma subordinator ($\alpha = 2$), geometric $\beta/2$-stable subordinators ($\alpha = \beta$), iterated geometric stable subordinators, Bessel subordinators $S_I$, $\alpha = 1$, and $S_K$, $\alpha = 2$, and also subordinators corresponding to Examples 5.15 and 5.16.

**Remark 5.42.** *Suppose that $S_t = bt + \tilde{S}_t$ where $b$ is positive and $\tilde{S}_t$ is a pure jump special subordinator with finite expectation. Then $\phi(\lambda) \sim b\lambda$, $\lambda \to \infty$, and $\phi(\lambda) \sim \phi'(0+)\lambda$, $\lambda \to 0$. This implies that, when $d \geq 3$, the Green function of the subordinate process $Y$ satisfies $G(x) \asymp G^{(2)}(x)$ for all $x \in \mathbb{R}^d$.*

### 5.3.3  Asymptotic Behavior of the Jumping Function

The goal of this subsection is to establish results on the asymptotic behavior
of the jumping function near zero, and results about the rate of decay of
the jumping function near zero and near infinity. We start by stating two
theorems on the asymptotic behavior of the jumping functions at zero for
subordinate Brownian motions via subordinators corresponding to Examples
5.15 and 5.16. We omit the proofs which rely on Lemma 5.32 and are similar
to proofs of Theorems 5.39 and 5.40.

**Theorem 5.43.** *Suppose* $\alpha \in (0,2), \beta \in (0, 2 - \alpha)$ *and that* $S$ *is the subordinator corresponding to Example 5.15. Then the jumping function of the
subordinate Brownian motion* $Y$ *via* $S$ *satisfies*

$$J(x) \sim \frac{\alpha \Gamma((d+\alpha)/2)}{2^{1-\alpha}\pi^{d/2}\Gamma(1-\alpha/2)} \frac{(\log(1/|x|^2))^{\beta/2}}{|x|^{d+\alpha}}, \quad |x| \to 0.$$

**Theorem 5.44.** *Suppose* $\alpha \in (0,2), \beta \in (0,\alpha)$ *and that* $S$ *is the subordinator
corresponding to Example 5.16. Then the jumping function of the subordinate
Brownian motion* $Y$ *via* $S$ *satisfies*

$$J(x) \sim \frac{\alpha \Gamma((d+\alpha)/2)}{2^{1-\alpha}\pi^{d/2}\Gamma(1-\alpha/2)} \frac{1}{|x|^{d+\alpha}(\log(1/|x|^2))^{\beta/2}}, \quad |x| \to 0.$$

We continue by establishing the asymptotic behavior of the jumping function for the geometric stable processes. More precisely, for $0 < \alpha \le 2$, let
$\phi(\lambda) = \log(1 + \lambda^{\alpha/2})$, $S$ the corresponding geometric $\alpha/2$-stable subordinator, $Y_t = X(S_t)$ the subordinate process and $J$ the jumping function of $Y$.

**Theorem 5.45.** *For every* $\alpha \in (0, 2]$, *it holds that*

$$J(x) \sim \frac{\alpha \Gamma(d/2)}{2|x|^d}, \quad |x| \to 0.$$

**Proof.** We again apply Lemma 5.32, this time with $w(t) = \mu(t)$, the density
of the Lévy measure of $S$. By Proposition 5.24 (i), $\mu(t) \sim \frac{\alpha}{2t}$ as $t \to 0+$, so
we take $c_0 = \alpha/2$, $\beta = 1$ and $\ell(t) = 1$. By Proposition 5.24 (ii), $\mu(t)$ is of the
order $t^{-\alpha/2-1}$ as $t \to +\infty$, so we may take $\gamma = -\alpha/2$. Choose $\xi = 1/2$ and
let $g = 1$.                                                                      □

**Theorem 5.46.** *For every* $\alpha \in (0,2)$ *we have*

$$J(x) \sim \frac{\alpha}{2^{\alpha+1}\pi^{d/2}} \frac{\Gamma(\frac{d+\alpha}{2})}{\Gamma(1-\frac{\alpha}{2})} |x|^{-d-\alpha}, \quad |x| \to \infty.$$

**Proof.** By Proposition 5.24 (ii),

$$\mu(t) \sim \frac{\alpha}{2\Gamma(1-\alpha/2)} t^{-\alpha/2-1}, \quad t \to \infty.$$

Now combine this with Proposition 5.24 (i) to get that

$$\mu(t) \le C(t^{-1} \vee t^{-\alpha/2-1}), \quad t > 0. \tag{5.60}$$

By change of variables we have

$$\int_0^\infty (4\pi t)^{-d/2} \exp\left(-\frac{|x|^2}{4t}\right) \mu(t)\, dt$$

$$= \frac{1}{4\pi^{d/2}} |x|^{-d+2} \int_0^\infty s^{d/2-2} e^{-s} \mu\left(\frac{|x|^2}{4s}\right) ds$$

$$= \frac{\alpha}{8\pi^{d/2}\Gamma(1-\alpha/2)} |x|^{-d-\alpha} \int_0^\infty s^{d/2-2} e^{-s} \frac{\mu\left(\frac{|x|^2}{4s}\right)}{\frac{\alpha}{2\Gamma(1-\alpha/2)}\left(\frac{|x|^2}{4s}\right)^{-\alpha/2-1}} \left(\frac{1}{4s}\right)^{-\alpha/2-1} ds$$

$$= \frac{\alpha}{2^{\alpha+1}\pi^{d/2}\Gamma(1-\alpha/2)} |x|^{-d-\alpha} \int_0^\infty s^{d/2+\alpha/2-1} e^{-s} \frac{\mu\left(\frac{|x|^2}{4s}\right)}{\frac{\alpha}{a\Gamma(1-\alpha/2)}\left(\frac{|x|^2}{4s}\right)^{-\alpha/2-1}} ds.$$

Let $|x| \ge 2$. Then by (5.60),

$$\frac{u\left(\frac{|x|^2}{4s}\right)}{\left(\frac{|x|^2}{4s}\right)^{-\alpha/2-1}} \le C\left(\left(\frac{|x|^2}{4s}\right)^{\alpha/2} \vee 1\right) \le C(s^{-\alpha/2} \vee 1).$$

It follows that the integrand in the last display above is bounded by an integrable function, so we may use the dominated convergence theorem to obtain

$$\lim_{|x|\to\infty} \frac{1}{|x|^{-d-\alpha}} \int_0^\infty (4\pi t)^{-d/2} \exp\left(-\frac{|x|^2}{4t}\right) \mu(t)\, dt = \frac{\alpha}{2^{\alpha+1}\pi^{d/2}} \frac{\Gamma(\frac{d+\alpha}{2})}{\Gamma(1-\frac{\alpha}{2})}, \tag{5.61}$$

which proves the result.                                                $\square$

In the case $\alpha = 2$, the behavior of $J$ at $\infty$ is different and is given in the following result.

**Theorem 5.47.** *When $\alpha = 2$, we have*

$$J(x) \sim 2^{-d/2} \pi^{-\frac{d-1}{2}} \frac{e^{-|x|}}{|x|^{\frac{d+1}{2}}}, \quad |x| \to \infty.$$

**Proof.** By change of variables we get that

$$J(x) = \frac{1}{2} \int_0^\infty t^{-1} e^{-t} (4\pi t)^{-d/2} \exp(-\frac{|x|^2}{2}) dt$$

$$= 2^{-d-1} \pi^{-d/2} |x|^{-d} \int_0^\infty s^{\frac{d}{2}-1} e^{-\frac{s}{4} - \frac{|x|^2}{s}} ds$$

$$= 2^{-d-1} \pi^{-d/2} |x|^{-d} I(|x|),$$

where

$$I(r) = \int_0^\infty s^{\frac{d}{2}-1} e^{-\frac{s}{4} - \frac{r^2}{s}} ds.$$

Using the change of variable $u = \frac{\sqrt{s}}{2} - \frac{r}{\sqrt{s}}$ we get

$$I(r) = e^{-r} \int_0^\infty s^{\frac{d}{2}-1} e^{-(\frac{\sqrt{s}}{2} - \frac{r}{\sqrt{s}})^2} ds$$

$$= e^{-r} \int_{-\infty}^\infty \frac{2(u + \sqrt{u^2 + 2r})^d}{\sqrt{u^2 + 2r}} e^{-u^2} du$$

$$= 2e^{-r} r^{\frac{d-1}{2}} \int_{-\infty}^\infty \frac{u + \sqrt{u^2 + 2r}}{\sqrt{u^2 + 2r}} (\frac{u}{\sqrt{r}} + \sqrt{\frac{u^2}{r} + 2})^{d-1} e^{-u^2} du.$$

Therefore by the dominated convergence theorem we obtain

$$I(r) \sim 2^{\frac{d}{2}+1} \sqrt{\pi} e^{-r} r^{\frac{d-1}{2}}, \quad r \to \infty.$$

Now the assertion of the theorem follows immediately.                    □

Let $Y_t^{(n)} = X(S_t^{(n)})$ be Brownian motion subordinated by the iterated geometric subordinator $S^{(n)}$, and let $J^{(n)}$ be the corresponding jumping function. Because of Remark 5.26, we were unable to determine the asymptotic behavior of $J^{(n)}$.

Assume now that $\phi(\lambda)$ is a complete Bernstein function which asymptotically behaves as $\lambda^{\alpha/2}$ as $\lambda \to 0+$ (resp. as $\lambda \to \infty$). Similar arguments as in Theorems 5.45 and 5.46 would yield that the jumping function $J$ of the corresponding subordinate Brownian motion behaves (up to a constant) as $|x|^{-\alpha-d}$ as $|x| \to \infty$ (resp. as $|x|^{-\alpha-d}$ as $|x| \to 0$). We are not going to pursue this here, because, firstly, such behavior of the jumping kernel is known from the case of $\alpha$-stable processes, and secondly, in the sequel we will not be interested in precise asymptotics of $J$, but rather in the rate of decay near zero and near infinity. Recall that $\mu(t)$ denotes the decreasing density of the Lévy measure of the subordinator $S$ (which exists since $\phi$ is assumed to be complete Bernstein), and recall that the function $j : (0, \infty) \to (0, \infty)$ was defined by

$$j(r) := \int_0^\infty (4\pi)^{-d/2} t^{-d/2} \exp\left(-\frac{r^2}{4t}\right) \mu(t)\, dt\,, \quad r > 0\,, \tag{5.62}$$

and that $J(x) = j(|x|)$, $x \in \mathbb{R}^d \setminus \{0\}$.

**Proposition 5.48.** *Suppose that there exists a positive constant $c_1 > 0$ such that*

$$\mu(t) \le c_1 \mu(2t) \quad \text{for all } t \in (0,8) \,, \tag{5.63}$$
$$\mu(t) \le c_1 \mu(t+1) \quad \text{for all } t > 1 \,. \tag{5.64}$$

*Then there exists a positive constant $c_2$ such that*

$$j(r) \le c_2 j(2r) \quad \text{for all } r \in (0,2) \,, \tag{5.65}$$
$$j(r) \le c_2 j(r+1) \quad \text{for all } r > 1 \,. \tag{5.66}$$

*Also, $r \mapsto j(r)$ is decreasing on $(0, \infty)$.*

**Proof.** For simplicity we redefine in this proof the function $j$ by dropping the factor $(4\pi)^{-d/2}$ from its definition. This does not effect (5.65) and (5.66).
  Let $0 < r < 2$. We have

$$j(2r) = \int_0^\infty t^{-d/2} \exp(-r^2/t)\mu(t)\, dt$$

$$= \frac{1}{2}\left( \int_0^{1/2} t^{-d/2} \exp(-r^2/t)\mu(t)\, dt + \int_{1/2}^\infty t^{-d/2} \exp(-r^2/t)\mu(t)\, dt \right.$$

$$\left. + \int_0^2 t^{-d/2} \exp(-r^2/t)\mu(t)\, dt + \int_2^\infty t^{-d/2} \exp(-r^2/t)\mu(t)\, dt \right)$$

$$\ge \frac{1}{2}\left( \int_{1/2}^\infty t^{-d/2} \exp(-r^2/t)\mu(t)\, dt + \int_0^2 t^{-d/2} \exp(-r^2/t)\mu(t)\, dt \right)$$

$$= \frac{1}{2}(I_1 + I_2).$$

Now,

$$I_1 = \int_{1/2}^\infty t^{-d/2} \exp(-\frac{r^2}{t})\mu(t)\, dt = \int_{1/2}^\infty t^{-d/2} \exp(-\frac{r^2}{4t}) \exp(-\frac{3r^2}{4t})\mu(t)\, dt$$

$$\ge \int_{1/2}^\infty t^{-d/2} \exp(-\frac{r^2}{4t}) \exp(-\frac{3r^2}{2})\mu(t)\, dt \ge e^{-6} \int_{1/2}^\infty t^{-d/2} \exp(-\frac{r^2}{4t})\mu(t)\, dt \,,$$

$$I_2 = \int_0^2 t^{-d/2} \exp(-\frac{r^2}{t})\mu(t)\, dt = 4^{-d/2+1} \int_0^{1/2} s^{-d/2} \exp(-\frac{r^2}{4s})\mu(4s)\, ds$$

$$\ge c_1^{-2} 4^{-d/2+1} \int_0^{1/2} s^{-d/2} \exp(-\frac{r^2}{4s})\mu(s)\, ds.$$

Combining the three displays above we get that $j(2r) \ge c_3\, j(r)$ for all $r \in (0,2)$.

To prove (5.66) we first note that for all $t \geq 2$ and all $r \geq 1$ it holds that

$$\frac{(r+1)^2}{t} - \frac{r^2}{t-1} \leq 1.$$

This implies that

$$\exp(-\frac{(r+1)^2}{4t}) \geq e^{-1/4} \exp(-\frac{r^2}{4(t-1)}), \qquad \text{for all } r > 1, t > 2. \qquad (5.67)$$

Now we have

$$j(r+1) = \int_0^\infty t^{-d/2} \exp(-\frac{(r+1)^2}{4t})\mu(t)\, dt$$

$$\geq \frac{1}{2} \left( \int_0^8 t^{-d/2} \exp(-\frac{(r+1)^2}{4t})\mu(t)\, dt + \int_3^\infty t^{-d/2} \exp(-\frac{(r+1)^2}{4t})\mu(t)\, dt \right)$$

$$= \frac{1}{2}(I_3 + I_4).$$

For $I_3$ note that $(r+1)^2 \leq 4r^2$ for all $r > 1$. Thus

$$I_3 = \int_0^8 t^{-d/2} \exp(-\frac{(r+1)^2}{4t})\mu(t)\, dt \geq \int_0^8 t^{-d/2} \exp(-r^2/t)\mu(t)\, dt$$

$$= 4^{-d/2+1} \int_0^2 s^{-d/2} \exp(-\frac{r^2}{4s})\mu(4s)\, ds$$

$$\geq c_1^{-2} 4^{-d/2+1} \int_0^2 s^{-d/2} \exp(-\frac{r^2}{4s})\mu(s)\, ds,$$

$$I_4 = \int_3^\infty t^{-d/2} \exp(-\frac{(r+1)^2}{4t})\mu(t)\, dt$$

$$\geq \int_3^\infty t^{-d/2} \exp\{-1/4\} \exp(-\frac{r^2}{4(t-1)})\,\mu(t)\, dt$$

$$= e^{-1/4} \int_2^\infty (s-1)^{-d/2} \exp(-\frac{r^2}{4s})\,\mu(s+1)\, ds$$

$$\geq c_1^{-1} e^{-1/4} \int_2^\infty s^{-d/2} \exp(-\frac{r^2}{4s})\mu(s)\, ds.$$

Combining the three displays above we get that $j(r+1) \geq c_4 j(r)$ for all $r > 1$.                                                                                              □

Suppose that $S = (S_t : t \geq 0)$ is an $\alpha/2$-stable subordinator, or a relativistic $\alpha/2$-stable subordinator, or a gamma subordinator. By the explicit forms of the Lévy densities given in Examples 5.8, 5.9 and 5.10 it is straightforward to verify that in all three cases $\mu(t)$ satisfies (5.63) and (5.64). For the Bessel subordinators, by use of asymptotic behavior of modified Bessel functions $I_0$ and $K_0$, one obtains that $\mu_I(t) \sim e^{-t}/t$, $t \to 0+$, $\mu_I(t) \sim (1/\sqrt{2\pi})\,t^{-3/2}$, $t \to \infty$, $\mu_K(t) \sim \log(1/t)/t$, $t \to 0+$, and $\mu_K(t) \sim \sqrt{\pi/2}\,e^{-2t}\,t^{-3/2}$, $t \to \infty$. From Propositions 5.27 and 5.28, it is easy to see that corresponding Lévy densities satisfy (5.63) and (5.64). In the case when $S$ is a geometric $\alpha/2$-stable subordinator or when $S$ is the subordinator corresponding to Example 5.15, respectively Example 5.16, these two properties follow from Proposition 5.24, and Proposition 5.27, respectively Proposition 5.28. In the case of an iterated geometric stable subordinator with $0 < \alpha < 2$, (5.64) is a consequence of Proposition 5.25, but we do not know whether (5.63) holds true. By using a different approach, we will show that if $j^{(n)} : (0, \infty) \to (0, \infty)$ is such that $J^{(n)}(x) = j^{(n)}(|x|)$, then (5.65) and (5.66) are still true.

We first observe that symmetric geometric $\alpha$-stable process $Y$ can be obtained by subordinating a symmetric $\alpha$-stable process $X^\alpha$ via a gamma subordinator $S$. Indeed, the characteristic exponent of $X^\alpha$ being equal to $|x|^\alpha$, and the Laplace exponent of $S$ being equal to $\log(1+\lambda)$, the composition of these two gives the characteristic exponent $\log(1 + |x|^\alpha)$ of a symmetric geometric $\alpha$-stable process. Let $p_\alpha(t, x, y) = p_\alpha(t, x - y)$ denote the transition densities of the symmetric $\alpha$-stable process, and let $q_\alpha(t, x, y) = q_\alpha(t, x - y)$ denote the transition densities of the symmetric geometric $\alpha$-stable process, $x, y \in \mathbb{R}^d$, $t \geq 0$. Then

$$q_\alpha(t, x) = \int_0^\infty p_\alpha(s, x) \frac{1}{\Gamma(t)} s^{t-1} e^{-s} ds. \tag{5.68}$$

Also, similarly as in (5.48), the jumping function of $Y$ can be written as

$$J(x) = \int_0^\infty p_\alpha(t, x) t^{-1} e^{-t}\, dt, \quad x \in \mathbb{R}^d \setminus \{0\}. \tag{5.69}$$

Define functions $j^{(n)} : (0, \infty) \to (0, \infty)$ by

$$j^{(n)}(r) := \int_0^\infty t^{-d/2} \exp\left(-\frac{r^2}{4t}\right) \mu^{(n)}(t)\, dt, \quad r > 0, \tag{5.70}$$

where $\mu^{(n)}$ denotes the Lévy density of the iterated geometric subordinator, and note that by (5.48), $J^{(n)}(x) = (4\pi)^{-d/2} j^{(n)}(|x|)$, $x \in \mathbb{R}^d \setminus \{0\}$.

**Proposition 5.49.** *For any $\alpha \in (0, 2)$ and $n \geq 1$, there exists a positive constant $c$ such that*

$$j^{(n)}(r) \leq c j^{(n)}(2r), \quad \text{for all } r > 0 \tag{5.71}$$

*and*

$$j^{(n)}(r) \leq c j^{(n)}(r + 1), \quad \text{for all } r > 1. \tag{5.72}$$

**Proof.** The inequality (5.72) follows from Proposition 5.48. Now we prove (5.71). It is known (see Theorem 2.1 of [26]) that there exist positive constants $C_1$ and $C_2$ such that for all $t > 0$ and all $x \in \mathbb{R}^d$,

$$C_1 \min(t^{-d/\alpha}, t\,|x|^{-d-\alpha}) \leq p_\alpha(t, x) \leq C_2 \min(t^{-d/\alpha}, t\,|x|^{-d-\alpha}). \quad (5.73)$$

Using these estimates one can easily see that there exists $C_3 > 0$ such that

$$p_\alpha(t, x) \leq C_3 p_\alpha(t, 2x), \quad \text{for all } t > 0 \text{ and } x \in \mathbb{R}^d. \quad (5.74)$$

Let $J^{(1)}(x) = J(x)$ and $q_\alpha^{(1)}(t, x) = q_\alpha(t, x)$. By use of (5.74), it follows from (5.68) and (5.69) that $J^{(1)}(x) \leq C_3 J^{(1)}(2x)$, for all $x \in \mathbb{R}^d \setminus \{0\}$, and $q_\alpha^{(1)}(t, x) \leq C_3 q_\alpha^{(1)}(t, 2x)$, for all $t > 0$ and $x \in \mathbb{R}^d$. Further, $Y^{(2)}$ is obtained by subordinating $Y^{(1)}$ by a geometric $\alpha/2$-stable subordinator $S$. Therefore,

$$J^{(2)}(x) = \frac{1}{2} \int_0^\infty q_\alpha^{(1)}(s, x) \mu_{\alpha/2}(s) \, ds, \quad q_\alpha^{(2)}(t, x) = \int_0^\infty q_\alpha^{(1)}(s, x) f_{\alpha/2}(t, s) \, ds,$$

$$(5.75)$$

where $\mu(s)$ is the Lévy density of $S$ and $f_{\alpha/2}(t, s)$ the density of $\mathbb{P}(S_t \in ds)$. By use of $q_\alpha^{(1)}(s, x) \leq C_3 q_\alpha^{(1)}(s, 2x)$, it follows $J^{(2)}(x) \leq C_3 J^{(2)}(2x)$ and $q_\alpha^{(2)}(t, x) \leq C_3 q_\alpha^{(2)}(t, 2x)$ for all $t > 0$ and $x \in \mathbb{R}^d$. The proof is completed by induction. $\qquad \square$

We conclude this section with a result that is essential in proving the Harnack inequality for jump processes, and was the motivation behind Propositions 5.48 and 5.49.

**Proposition 5.50.** *Let $Y$ be a subordinate Brownian motion such that the function $j$ defined in (5.62) satisfies conditions (5.65) and (5.66). There exist positive constants $C_4$ and $C_5$ such that if $r \in (0, 1)$, $x \in B(0, r)$, and $H$ is a nonnegative function with support in $B(0, 2r)^c$, then*

$$\mathbb{E}^x H(Y(\tau_{B(0,r)})) \leq C_4 (\mathbb{E}^x \tau_{B(0,r)}) \int H(z) J(z) \, dz$$

*and*

$$\mathbb{E}^x H(Y(\tau_{B(0,r)})) \geq C_5 (\mathbb{E}^x \tau_{B(0,r)}) \int H(z) J(z) \, dz.$$

**Proof.** Let $y \in B(0, r)$ and $z \in B(0, 2r)^c$. If $z \in B(0, 2)$ we use the estimates

$$2^{-1}|z| \leq |z - y| \leq 2|z|, \quad (5.76)$$

while if $z \notin B(0, 2)$ we use

$$|z| - 1 \leq |z - y| \leq |z| + 1. \quad (5.77)$$

Let $B \subset B(0, 2r)^c$. Then by using the Lévy system we get

$$\mathbb{E}^x 1_B(Y(\tau_{B(0,r)})) = \mathbb{E}^x \int_0^{\tau_{B(0,r)}} \int_B J(z - Y_s)\, dz\, ds$$

$$= \mathbb{E}^x \int_0^{\tau_{B(0,r)}} \int_B j(|z - Y_s|)\, dz\, ds\,.$$

By use of (5.65), (5.66), (5.76), and (5.77), the inner integral is estimated as follows:

$$\int_B j(|z - Y_s|)\, dz = \int_{B \cap B(0,2)} j(|z - Y_s|)\, dz + \int_{B \cap B(0,2)^c} j(|z - Y_s|)\, dz$$

$$\leq \int_{B \cap B(0,2)} j(2^{-1}|z|)\, dz + \int_{B \cap B(0,2)^c} j(|z| - 1)\, dz$$

$$\leq \int_{B \cap B(0,2)} c_2 j(|z|)\, dz + \int_{B \cap B(0,2)^c} c_2 j(|z|)\, dz$$

$$= c_2 \int_B J(z)\, dz\,.$$

Therefore

$$\mathbb{E}^x 1_B(Y(\tau_{B(0,r)})) \leq \mathbb{E}^x \int_0^{\tau_{B(0,r)}} c_2 \int_B J(z)\, dz$$

$$= c_2\, \mathbb{E}^x(\tau_{B(0,r)}) \int 1_B(z) J(z)\, dz\,.$$

Using linearity we get the above inequality when $1_B$ is replaced by a simple function. Approximating $H$ by simple functions and taking limits we have the first inequality in the statement of the lemma.

The second inequality is proved in the same way.                    $\square$

### 5.3.4  Transition Densities of Symmetric Geometric Stable Processes

Recall that for $0 < \alpha \leq 2$, $q_\alpha(t, x)$ denotes the transition density of the symmetric geometric $\alpha$-stable process. The asymptotic behavior of $q_\alpha(1, x)$ as $|x| \to \infty$ is given in the following result.

**Proposition 5.51.** *For* $\alpha \in (0, 2)$ *we have*

$$q_\alpha(1, x) \sim \frac{\alpha 2^{\alpha-1} \sin \frac{\alpha\pi}{2} \Gamma(\frac{d+\alpha}{2}) \Gamma(\frac{\alpha}{2})}{\pi^{\frac{d}{2}+1} |x|^{d+\alpha}}, \quad |x| \to \infty.$$

*For $\alpha = 2$ we have*

$$q_2(1, x) \sim 2^{-\frac{d}{2}} \pi^{-\frac{d-1}{2}} \frac{e^{-|x|}}{|x|^{\frac{d-1}{2}}}, \qquad |x| \to \infty.$$

**Proof.** The proof of the case $\alpha < 2$ is similar to the proof of Proposition 5.29 and uses (5.73), while the proof of the case $\alpha = 2$ is similar to the proof of Theorem 5.47. We omit the details. □

The following theorem from [59] provides the sharp estimate for $q_\alpha(t, x)$ for small time $t$ in case $0 < \alpha < 2$.

**Theorem 5.52.** *Let $\alpha \in (0, 2)$. There are positive constants $C_1 < C_2$ such that for all $x \in \mathbb{R}^d$ and $0 < t < 1 \wedge \frac{d}{2\alpha}$,*

$$C_1 t \min(|x|^{-d-\alpha}, |x|^{-d+t\alpha}) \le q_\alpha(t, x) \le C_2 t \min(|x|^{-d-\alpha}, |x|^{-d+t\alpha}).$$

**Proof.** The following sharp estimates for the stable densities (5.73) is well known (see, for instance, [26])

$$p_\alpha(s, x) \asymp s^{-\frac{d}{\alpha}} \left( 1 \wedge \frac{s^{\frac{d+\alpha}{\alpha}}}{|x|^{d+\alpha}} \right), \qquad \forall s > 0 \text{ and } x \in \mathbb{R}^d.$$

Hence, by (5.68) it follows that $q_\alpha(t, x) \asymp \frac{1}{\Gamma(t)} I(t, |x|)$ where

$$I(t, r) := \int_0^\infty s^{-\frac{d}{\alpha}} \left( 1 \wedge \frac{s^{\frac{d+\alpha}{\alpha}}}{r^{d+\alpha}} \right) s^{t-1} e^{-s} \, ds$$

$$= \frac{1}{r^{d+\alpha}} \int_0^{r^\alpha} s^t e^{-s} \, ds + \int_{r^\alpha}^\infty s^{t-1-d/\alpha} e^{-s} \, ds.$$

From now on assume that $0 < t \le 1 \wedge \frac{d}{2\alpha}$. Then for $0 < r \le 1$,

$$I(t, r) \asymp \frac{1}{r^{d+\alpha}} \int_0^{r^\alpha} s^t \, ds + \int_{r^\alpha}^1 s^{t-1-d/\alpha} e^{-s} \, ds + \int_1^\infty s^{t-1-d/\alpha} e^{-s} \, ds$$

$$= \frac{1}{t+1} r^{\alpha t - d} + \frac{1}{d/\alpha - t} (r^{\alpha t - d} - 1) + \int_1^\infty s^{t-1-d/\alpha} e^{-s} \, ds \asymp r^{\alpha t - d}.$$

We also have,

$$I(t, r) \le \frac{1}{r^{d+\alpha}} \int_0^\infty s^t e^{-s} \, ds = \frac{\Gamma(t+1)}{r^{d+\alpha}} \le \frac{1}{r^{d+\alpha}}, \qquad r > 0,$$

$$I(t, r) \ge \frac{1}{r^{d+\alpha}} \int_0^1 s^t e^{-s} \, ds \ge \frac{1}{r^{d+\alpha}} \frac{1}{(1+t)e} \ge \frac{1}{2e r^{d+\alpha}}, \qquad r > 1.$$

Note that

$$\frac{1}{r^{d+\alpha}} \wedge \frac{1}{r^{d-t\alpha}} = \begin{cases} \frac{1}{r^{d-t\alpha}}, & 0 < r \le 1 \\ \frac{1}{r^{d+\alpha}}, & r > 1. \end{cases}$$

Therefore $I(t,r) \asymp \frac{1}{r^{d+\alpha}} \wedge \frac{1}{r^{d-t\alpha}}$. This implies that

$$q_\alpha(t,x) \asymp \frac{1}{\Gamma(t)} I(t,|x|) \asymp \frac{1}{\Gamma(t)} \left( \frac{1}{|x|^{d+\alpha}} \wedge \frac{1}{|x|^{d-t\alpha}} \right) \asymp t \left( \frac{1}{|x|^{d+\alpha}} \wedge \frac{1}{|x|^{d-t\alpha}} \right),$$

since for $0 < t \le 1$, $\Gamma(t) \asymp t^{-1}$.                                     $\square$

Note that by taking $x = 0$, one obtains that $q_\alpha(t,0) = \infty$ for $0 < t < 1 \wedge \frac{d}{2\alpha}$. This somewhat unusual feature of the transition density is easier to show when $\alpha = 2$, i.e., in the case of a gamma subordinator. Indeed, then

$$q_2(t,x) := \int_0^\infty (4\pi s)^{-d/2} e^{-|x|^2/(4t)} \frac{1}{\Gamma(t)} s^{t-1} e^{-s} \, ds \,,$$

and therefore

$$q_2(t,0) = \frac{(4\pi)^{-d/2}}{\Gamma(t)} \int_0^\infty s^{-d/2+t-1} e^{-s} \, ds = \begin{cases} +\infty\,, & t \le d/2, \\ \frac{\Gamma(t-d/2)}{(4\pi)^{d/2}\Gamma(t)}\,, & t > d/2\,. \end{cases}$$

Assume now that $S^{(2)}$ is an iterated geometric stable subordinator with the Laplace exponent $\phi(\lambda) = \log(1 + \log(1 + \lambda))$, and let $q_2^{(2)}(t,x)$ be the transition density of the process $Y_t^{(2)} = X(S_t^{(2)})$. Then by (5.75),

$$q_2^{(2)}(t,0) = \int_0^\infty q_2(s,0) \frac{1}{\Gamma(t)} s^{t-1} e^{-s} \, ds = \infty$$

for all $t > 0$.

## 5.4   Harnack Inequality for Subordinate Brownian Motion

### 5.4.1   Capacity and Exit Time Estimates for Some Symmetric Lévy Processes

The purpose of this subsection is to establish lower and upper estimates for the capacity of balls and the exit time from balls, with respect to a class of radially symmetric Lévy processes.

Suppose that $Y = (Y_t, \mathbb{P}^x)$ is a transient radially symmetric Lévy process on $\mathbb{R}^d$. We will assume that the potential kernel of $Y$ is absolutely continuous with a density $G(x, y) = G(|y - x|)$ with respect to the Lebesgue measure. Let us assume the following condition: $G : [0, \infty) \to (0, \infty]$ *is a positive and decreasing function satisfying* $G(0) = \infty$. We will have need of the following elementary lemma.

**Lemma 5.53.** *There exists a positive constant* $C_1 = C_1(d)$ *such that for every* $r > 0$ *and all* $x \in \overline{B(0, r)}$,

$$C_1 \int_{B(0,r)} G(|y|) \, dy \le \int_{B(0,r)} G(x, y) \, dy \le \int_{B(0,r)} G(|y|) \, dy \, .$$

*Moreover, the supremum of* $\int_{B(0,r)} G(x, y) \, dy$ *is attained at* $x = 0$, *while the infimum is attained at any point on the boundary of* $B(0, r)$.

**Proof.** The proof is elementary. We only present the proof of the left-hand side inequality for $d \ge 2$. Consider the intersection of $B(0, r)$ and $B(x, r)$. This intersection contains the intersection of $B(x, r)$ and the cone with vertex $x$ of aperture equal to $\pi/3$ pointing towards the origin. Let $C(x)$ be the latter intersection. Then

$$\int_{B(0,r)} G(|y - x|) \, dy \ge \int_{C(x)} G(|y - x|) \, dy$$

$$\ge c_1 \int_{B(x,r)} G(|y - x|) \, dy = c_1 \int_{B(0,r)} G(|y|) \, dy \, ,$$

where the constant $c_1$ depends only on the dimension $d$. It is easy to see that the infimum of $\int_{B(0,r)} G(x, y) \, dy$ is attained at any point on the boundary of $B(0, r)$.                                                                      □

Let Cap denote the (0-order) capacity with respect to $X$ (for the definition of capacity see e.g. [23] or [135]). For a measure $\mu$ we define

$$G\mu(x) := \int G(x, y) \, \mu(dy) \, .$$

For any compact subset $K$ of $\mathbb{R}^d$, let $\mathcal{P}_K$ be the set of probability measures supported by $K$. Define

$$e(K) := \inf_{\mu \in \mathcal{P}_K} \int G\mu(x) \, \mu(dx) \, .$$

Since the kernel $G$ satisfies the maximum principle (see, for example, Theorem 5.2.2 in [60]), it follows from ([76], page 159) that for any compact subset $K$ of $\mathbb{R}^d$

$$\text{Cap}(K) = \frac{1}{\inf_{\mu \in \mathcal{P}_K} \sup_{x \in \text{Supp}(\mu)} G\mu(x)} = \frac{1}{e(K)} \, . \tag{5.78}$$

Furthermore, the infimum is attained at the capacitary measure $\mu_K$. The following lemma is essentially proved in [114].

**Lemma 5.54.** *Let $K$ be a compact subset of $\mathbb{R}^d$. For any probability measure $\mu$ on $K$, it holds that*

$$\inf_{x \in \text{Supp}(\mu)} G\mu(x) \le e(K) \le \sup_{x \in \text{Supp}(\mu)} G\mu(x). \tag{5.79}$$

**Proof.** The right-hand side inequality follows immediately from (5.78). In order to prove the left-hand side inequality, suppose that for some probability measure $\mu$ on $K$ it holds that $e(K) < \inf_{x \in \text{Supp}(\mu)} G\mu(x)$. Then $e(K) + \epsilon < \inf_{x \in \text{Supp}(\mu)} G\mu(x)$ for some $\epsilon > 0$. We first have

$$\int_K G\mu(x)\, \mu_K(dx) > \int_K (e(K) + \epsilon)\, \mu_K(dx) = e(K) + \epsilon.$$

On the other hand,

$$\int_K G\mu(x)\, \mu_K(dx) = \int_K G\mu_K(x)\, \mu(dx) = \int_K e(K)\, \mu(dx) = e(K),$$

where we have used the fact that $G\mu_K = e(K)$ quasi everywhere in $K$, and the measure of finite energy does not charge sets of capacity zero. This contradiction proves the lemma. ☐

**Proposition 5.55.** *There exist positive constants $C_2 < C_3$ depending only on $d$, such that for all $r > 0$*

$$\frac{C_2 r^d}{\int_{B(0,r)} G(|y|)\, dy} \le \text{Cap}(\overline{B(0,r)}) \le \frac{C_3 r^d}{\int_{B(0,r)} G(|y|)\, dy}. \tag{5.80}$$

**Proof.** Let $m_r(dy)$ be the normalized Lebesgue measure on $B(0,r)$. Thus, $m_r(dy) = dy/(c_1 r^d)$, where $c_1$ is the volume of the unit ball. Consider $Gm_r = \sup_{x \in B(0,r)} Gm_r(x)$. By Lemma 5.53, the supremum is attained at $x = 0$, and so

$$Gm_r = \frac{1}{c_1 r^d} \int_{B(0,r)} G(|y|) dy.$$

Therefore from Lemma 5.54

$$\text{Cap}(\overline{B(0,r)}) \ge \frac{c_1 r^d}{\int_{B(0,r)} G(|y|) dy}. \tag{5.81}$$

For the right-hand side of (5.79), it follows from Lemma 5.53 and Lemma 5.54 that

$$\text{Cap}(\overline{B(0,r)}) \le \frac{1}{Gm_r(z)} = \frac{c_1 r^d}{\int_{B(0,r)} G(z,y)} dy \le \frac{c_1 r^d}{C_1 \int_{B(0,r)} G(|y|)} dy,$$

where $z \in \partial B(0,r)$. ☐

In the remaining part of this section we assume in addition that $G$ satisfies the following assumption: There exist $r_0 > 0$ and $c_0 \in (0,1)$ such that

$$c_0 G(r) \geq G(2r), \quad 0 < 2r < r_0. \qquad (5.82)$$

Note that if $G$ is regularly varying at 0 with index $\delta < 0$, i.e., if

$$\lim_{r \to 0} \frac{G(2r)}{G(r)} = 2^\delta,$$

then (5.82) is satisfied with $c_0 = (2^\delta + 1)/2$ for some positive $r_0$. Let $\tau_{B(0,r)} = \inf\{t > 0 : Y_t \notin B(0,r)\}$ be the first exit time of $Y$ from the ball $B(0,r)$.

**Proposition 5.56.** *There exists a positive constant $C_4$ such that for all $r \in (0, r_0/2)$,*

$$C_4 \int_{B(0,r/6)} G(|y|)\, dy \leq \inf_{x \in B(0,r/6)} \mathbb{E}^x \tau_{B(0,r)} \leq \sup_{x \in B(0,r)} \mathbb{E}^x \tau_{B(0,r)}$$

$$\leq \int_{B(0,r)} G(|y|)\, dy. \quad (5.83)$$

**Proof.** Let $G_{B(0,r)}(x,y)$ denote the Green function of the process $Y$ killed upon exiting $B(0,r)$. Clearly, $G_{B(0,r)}(x,y) \leq G(x,y)$, for $x, y \in B(0,r)$. Therefore,

$$\mathbb{E}^x \tau_{B(0,r)} = \int_{B(0,r)} G_{B(0,r)}(x,y)\, dy$$

$$\leq \int_{B(0,r)} G(x,y)\, dy \leq \int_{B(0,r)} G(|y|)\, dy.$$

For the left-hand side inequality, let $r \in (0, r_0/2)$, and let $x, y \in B(0,r/6)$. Then,

$$G_{B(0,r)}(x,y) = G(x,y) - \mathbb{E}^x G(Y(\tau_{B(0,r)}), y)$$

$$\geq G(|y-x|) - G(2|y-x|).$$

The last inequality follows because $|y - Y(\tau_{B(0,r)})| \geq \frac{2}{3}r \geq 2|y-x|$. Let $c_1 = 1 - c_0 \in (0,1)$. By (5.82) we have that for all $u \in (0, r_0)$, $G(u) - G(2u) \geq c_1 G(u)$. Hence, $G(|y-x|) - G(2|y-x|) \geq c_1 G(|y-x|)$, which implies that $G_{B(0,r)}(x,y) \geq c_1 G(x,y)$ for all $x, y \in B(0,r/6)$. Now, for $x \in B(0,r/6)$,

$$\mathbb{E}^x \tau_{B(0,r)} = \int_{B(0,r)} G_{B(0,r)}(x,y)\, dy \geq \int_{B(0,r/6)} G_{B(0,r)}(x,y)\, dy$$

$$\geq c_1 \int_{B(0,r/6)} G(x,y)\, dy \geq c_1 C_1 \int_{B(0,r/6)} G(|y|)\, dy,$$

where the last inequality follows from Lemma 5.53. $\qquad \square$

**Example 5.57.** We illustrate the last two propositions by applying them to the iterated geometric stable process $Y^{(n)}$ introduced in Example 5.31 (iv) and (v). Hence, we assume that $d > 2(\alpha/2)^n$. By a slight abuse of notation we define a function $G^{(n)} : [0, \infty) \to (0, \infty]$ by $G^{(n)}(|x|) = G^{(n)}(x)$. Note that by Theorem 5.37, $G$ is regularly varying at zero with index $\beta = -d$. Let $r_0$ be the constant from (5.82). Let us first look at the asymptotic behavior of $\int_{B(0,r)} G^{(n)}(|y|)\,dy$ for small $r$. We have

$$
\int_{B(0,r)} G^{(n)}(|y|)\,dy = c_d \int_0^r u^{d-1} G^{(n)}(u)\,du
$$

$$
\sim \frac{c_d \Gamma(d/2)}{\alpha \pi^{d/2}} \int_0^r \frac{u^{d-1}\,du}{u^d L_{n-1}(1/u^2) l_n^2(1/u^2)} = \frac{c_d \Gamma(d/2)}{2\alpha \pi^{d/2}} \int_0^{r^2} \frac{dv}{v L_{n-1}(1/v) l_n^2(1/v)}
$$

$$
= \frac{c_d \Gamma(d/2)}{2\alpha \pi^{d/2}} \frac{1}{l_n(1/r^2)} \sim c_{\alpha,d} \frac{1}{l_n(1/r)}, \quad r \to 0.
$$

It follows from Proposition 5.55 that there exist positive constants $C_5 \le C_6$ such that for all $r \in (0, 1/e_n)$,

$$
C_5 r^d l_n(1/r) \le \mathrm{Cap}(\overline{B(0,r)}) \le C_6 r^d l_n(1/r).
$$

Similarly, it follows from Proposition 5.56 that there exist positive constants $C_7 \le C_8$ such that for all $r \in (0, (1/e_n) \wedge (r_0/2))$,

$$
\frac{C_7}{l_n(1/r)} \le \inf_{x \in B(0,r/6)} \mathbb{E}^x \tau_{B(0,r)} \le \sup_{x \in B(0,r)} \mathbb{E}^x \tau_{B(0,r)} \le \frac{C_8}{l_n(1/r)}. \tag{5.84}
$$

Here we also used the fact that $l_n$ is slowly varying.

By use of Theorem 5.41 and Proposition 5.55, we can estimate capacity of large balls. It easily follows that as $r \to \infty$, $\mathrm{Cap}(\overline{B(0,r)})$ is of the order $r^{\alpha(\alpha/2)^{n-1}}$.

## 5.4.2 Krylov-Safonov-type Estimate

In this subsection we retain the assumptions from the beginning of the previous one. Thus, $Y = (Y_t, \mathbb{P}^x)$ is a transient radially symmetric Lévy process on $\mathbb{R}^d$ with the potential kernel having the density $G(x, y) = G(|y - x|)$ which is positive, decreasing and $G(0) = \infty$. Let $r_1 \in (0, 1)$ and let $\ell : (1/r_1, \infty) \to (0, \infty)$ be a slowly varying function at $\infty$. Let $\beta \in [0, 1]$ be such that $d + 2\beta - 2 > 0$. We introduce the following additional assumption about the density $G$: There exists a positive constant $c_1$ such that

$$
G(x) \sim \frac{c_1}{|x|^{d+2\beta-2}\ell(1/|x|^2)}, \quad |x| \to 0. \tag{5.85}
$$

If we abuse notation and let $G(|x|) = G(x)$, then $G$ is regularly varying at $0$ with index $-d - 2\beta + 2 < 0$, hence satisfies the assumption (5.82) with some $r_0 > 0$. In order to simplify notations, we define the function $g : (0, r_1) \to (0, \infty)$ by

$$g(r) = \frac{1}{r^{d+2\beta-2}\ell(1/r^2)}.$$

Clearly, $g$ is regularly varying at $0$ with index $-d - 2\beta + 2 < 0$. Let $\bar{g}$ be a monotone equivalent of $g$ at $0$. More precisely, we define $\bar{g} : (0, r_1/2) \to \infty$ by

$$\bar{g}(r) := \sup\{g(\rho) : r \le \rho \le r_1\}.$$

By the 0-version of Theorem 1.5.3. in [21], $\bar{g}(r) \sim g(r)$ as $r \to 0$. Moreover, $\bar{g}(r) \ge g(r)$, and $\bar{g}$ is decreasing. Let $r_2 = \min(r_0, r_1)$. There exist positive constants $C_9 < C_{10}$ such that

$$C_9 \bar{g}(r) \le G(r) \le C_{10} \bar{g}(r), \quad r < r_2. \tag{5.86}$$

We define

$$c = \max\left\{\frac{1}{3}\left(\frac{4C_{10}}{C_9}\right)^{\frac{1}{d+2\beta-2}}, 1\right\}. \tag{5.87}$$

Since $\bar{g}$ is regularly varying at $0$ with index $-d - 2\beta + 2$, there exists $r_3 > 0$ such that

$$\frac{1}{2}\left(\frac{1}{3c}\right)^{d+2\beta-2} \le \frac{\bar{g}(6cr)}{\bar{g}(2r)} \le 2\left(\frac{1}{3c}\right)^{d+2\beta-2}, \quad r < r_3. \tag{5.88}$$

Finally, let

$$R = \min(r_2, r_3, 1) = \min(r_0, r_1, r_3, 1). \tag{5.89}$$

**Lemma 5.58.** *There exists $C_{11} > 0$ such that for any $r \in (0, (7c)^{-1}R)$, any closed subset $A$ of $B(0, r)$, and any $y \in B(0, r)$,*

$$\mathbb{P}^y(T_A < \tau_{B(0,7cr)}) \ge C_{11} \bar{\kappa}(r) \frac{\mathrm{Cap}(A)}{\mathrm{Cap}(\overline{B(0, r)})},$$

*where*

$$\bar{\kappa}(r) = \frac{r^d \bar{g}(r)}{\int_0^r \rho^{d-1} \bar{g}(\rho)\, d\rho}. \tag{5.90}$$

**Proof.** Without loss of generality we may assume that $\mathrm{Cap}(A) > 0$. Let $G_{B(0,7cr)}$ be the Green function of the process obtained by killing $Y$ upon exiting from $B(0, 7cr)$. If $\nu$ is the capacitary measure of $A$ with respect to $Y$, then we have for all $y \in B(0, r)$,

$$G_{B(0,7cr)}\nu(y) = \mathbb{E}^y[G_{B(0,7cr)}\nu(Y_{T_A}) : T_A < \tau_{B(0,7cr)}]$$
$$\leq \sup_{z \in \mathbb{R}^d} G_{B(0,7cr)}\nu(z)\mathbb{P}^y(T_A < \tau_{B(0,7cr)})$$
$$\leq \mathbb{P}^y(T_A < \tau_{B(0,7cr)}).$$

On the other hand we have for all $y \in B(0,r)$,

$$G_{B(0,7cr)}\nu(y) = \int G_{B(0,7cr)}(y,z)\nu(dz) \geq \nu(A) \inf_{z \in B(0,r)} G_{B(0,7cr)}(y,z)$$
$$= \mathrm{Cap}(A) \inf_{z \in B(0,r)} G_{B(0,7cr)}(y,z).$$

In order to estimate the infimum in the last display, note that $G_{B(0,7cr)}(y,z) = G(y,z) - \mathbb{E}^y[G(Y_{\tau_{B(0,7cr)}},z)]$. Since $|y - z| < 2r < R$, it follows by (5.86) and the monotonicity of $\overline{g}$ that

$$G(y,z) \geq C_9\overline{g}(|z - y|) \geq C_9\overline{g}(2r). \tag{5.91}$$

Now we consider $G(Y_{\tau_{B(0,7cr)}},z)$. First note that $|Y_{\tau_{B(0,7cr)}} - z| \geq 7cr - r \geq 7cr - cr \geq 6cr$. If $|Y_{\tau_{B(0,7cr)}} - z| \leq R$, then by (5.86) and the monotonicity of $\overline{g}$,

$$G(Y_{\tau_{B(0,7cr)}},z) \leq C_{10}\overline{g}(|z - Y_{\tau_{B(0,7cr)}}|) \leq C_{10}\overline{g}(6cr).$$

If, on the other hand, $|Y_{\tau_{B(0,7cr)}} - z| \geq R$, then $G(Y_{\tau_{B(0,7cr)}},z) \leq G(w)$, where $w \in \mathbb{R}^d$ is any point such that $|w| = R$. Here we have used the monotonicity of $G$. For $|w| = R$ we have that $G(w) \leq C_{10}\overline{g}(|w|) = C_{10}\overline{g}(R) \leq C_{10}\overline{g}(6cr)$. Therefore

$$\mathbb{E}^y[G(Y_{\tau_{B(0,7cr)}},z)] \leq C_{10}\overline{g}(6cr). \tag{5.92}$$

By use of (5.91) and (5.92) we obtain

$$G_{B(0,7cr)}(y,z) \geq C_9\overline{g}(2r) - C_{10}\overline{g}(6cr)$$
$$= \overline{g}(2r)\left(C_9 - C_{10}\frac{\overline{g}(6cr)}{\overline{g}(2r)}\right)$$
$$\geq \overline{g}(2r)\left(C_9 - 2C_{10}\left(\frac{1}{3c}\right)^{d+2\beta-2}\right)$$
$$\geq \overline{g}(2r)\left(C_9 - 2C_{10}\frac{C_9}{4C_{10}}\right) = \frac{C_9}{2}\overline{g}(2r),$$

where the next to last line follows from (5.88) and the last from definition (5.87). By using one more time that $\overline{g}$ is regularly varying at 0, we conclude that there exists a constant $C_{12} > 0$ such that for all $y, z \in B(0,r)$,

$$G_{B(0,7cr)}(y,z) \geq C_{12}\overline{g}(r).$$

Further, it follows from Proposition 5.55 that there exists a constant $C_{13} > 0$, such that

$$\frac{C_{13}}{\text{Cap}(\overline{B(0,r)})} \frac{r^d}{\int_0^r \rho^{d-1} \overline{g}(\rho) \, d\rho} \leq 1. \tag{5.93}$$

Hence

$$G_{B(0,7cr)}(y,z) \geq C_{12} C_{13} \frac{1}{\text{Cap}(\overline{B(0,r)})} \frac{r^d \overline{g}(r)}{\int_0^r \rho^{d-1} \overline{g}(\rho) \, d\rho}$$

$$\geq C_{14} \frac{1}{\text{Cap}(\overline{B(0,r)})} \overline{\kappa}(r).$$

To finish the proof, note that

$$\mathbb{P}^y(T_A < \tau_{B(0,7cr)}) \geq G_{B(0,7cr)} \nu(y) \geq C_{14} \overline{\kappa}(r) \frac{\text{Cap}(A)}{\text{Cap}(\overline{B(0,r)})}.$$

$\square$

**Remark 5.59.** *Note that in the estimate (5.93) we could use $g$ instead of $\overline{g}$. Together with the fact that $\overline{g}(r) \geq g(r)$ this would lead to the hitting time estimate*

$$\mathbb{P}^y(T_A < \tau_{B(0,7cr)}) \geq C_{11} \kappa(r) \frac{\text{Cap}(A)}{\text{Cap}(\overline{B(0,r)})},$$

*where*

$$\kappa(r) = \frac{r^d g(r)}{\int_0^r \rho^{d-1} g(\rho) \, d\rho}. \tag{5.94}$$

We will apply the above lemma to subordinate Brownian motions. Assume, first, that $Y_t = X(S_t)$ where $S = (S_t : t \geq 0)$ is the special subordinator with the Laplace exponent $\phi$ satisfying $\phi(\lambda) \sim \lambda^{\alpha/2} \ell(\lambda)$, $\lambda \to \infty$, where $0 < \alpha < 2 \wedge d$, and $\ell$ is slowly varying at $\infty$. Then the Green function of $Y$ satisfies all assumptions of this subsection, in particular (5.85) with $\beta = 1 - \alpha/2$, see (5.28) and Lemma (5.32). Define $c$ as in (5.87) for appropriate $C_9$ and $C_{10}$ and $\beta = 1 - \alpha/2$, and let $R$ be as in (5.89).

**Proposition 5.60.** *Assume that $Y_t = X(S_t)$ where $S = (S_t : t \geq 0)$ is the special subordinator with the Laplace exponent $\phi$ satisfying one of the following two conditions: (i) $\phi(\lambda) \sim \lambda^{\alpha/2} \ell(\lambda)$, $\lambda \to \infty$, where $0 < \alpha < 2$, and $\ell$ is slowly varying at $\infty$, or (ii) $\phi(\lambda) \sim \lambda$, $\lambda \to \infty$. If $Y$ is transient, the following statements are true:*
*(a) There exists a constant $C_{15} > 0$ such that for any $r \in (0, (7c)^{-1}R)$, any closed subset $A$ of $B(0,r)$, and any $y \in B(0,r)$,*

$$\mathbb{P}^y(T_A < \tau_{B(0,7cr)}) \geq C_{15} \frac{\text{Cap}(A)}{\text{Cap}(\overline{B(0,r)})}.$$

*(b) There exists a constant $C_{16} > 0$ such that for any $r \in (0, R)$ we have*

$$\sup_{y \in B(0,r)} \mathbb{E}^y \tau_{B(0,r)} \le C_{16} \inf_{y \in B(0,r/6)} \mathbb{E}^y \tau_{B(0,r)} \,.$$

**Proof.** We give the proof for case (i), case (ii) being simpler.
(a) It suffices to show that $\overline{\kappa}(r)$, $r < (7c)^{-1}R$, is bounded from below by a positive constant. Note that $\overline{g}$ is regularly varying at 0 with index $-d + \alpha$. Hence there is a slowly varying function $\overline{\ell}$ such that $\overline{g}(r) = r^{-d+\alpha}\overline{\ell}(r)$. By Karamata's monotone density theorem one can conclude that

$$\int_0^r \rho^{d-1}\overline{g}(\rho)\,d\rho = \int_0^r \rho^{\alpha-1}\overline{\ell}(\rho)\,d\rho \sim \frac{1}{\alpha} r^\alpha \overline{\ell}(r) = \frac{1}{\alpha} r^d \overline{g}(r)\,, \quad r \to 0\,.$$

Therefore,

$$\overline{\kappa}(r) = \frac{r^d \overline{g}(r)}{\int_0^r \rho^{d-1}\overline{g}(\rho)\,d\rho} \sim \frac{1}{\alpha}\,.$$

(b) By Proposition 5.56 it suffices to show that $\int_{B(0,r)} G(|y|)\,dy \le c \int_{B(0,r/6)} G(|y|)\,dy$ for some positive constant $c$. But, by the proof of part (a), $\int_{B(0,r)} G(|y|)\,dy \asymp r^d \overline{g}(r)$, while $\int_{B(0,r/6)} G(|y|)\,dy \asymp (r/6)^d \overline{g}(r/6)$. Since $\overline{g}$ is regularly varying, the claim follows.                                    $\square$

**Proposition 5.61.** *Let $S^{(n)}$ be the iterated geometric stable subordinator and let $Y_t^{(n)} = X(S_t^{(n)})$ be the corresponding subordinate process. Assume that $d > 2(\alpha/2)^n$.*
*(a) Let $\gamma > 0$. There exists a constant $C_{17} > 0$ such that for any $r \in (0, (7c)^{-1}R)$, any closed subset $A$ of $B(0,r)$, and any $y \in B(0,r)$*

$$\mathbb{P}^y(T_A < \tau_{B(0,7cr)}) \ge C_{17}\, r^\gamma \frac{\mathrm{Cap}(A)}{\mathrm{Cap}(B(0,r))}\,.$$

*(b) There exists a constant $C_{18} > 0$ such that for any $r \in (0, R)$ we have*

$$\sup_{y \in B(0,r)} \mathbb{E}^y \tau_{B(0,r)} \le C_{18} \inf_{y \in B(0,r/6)} \mathbb{E}^y \tau_{B(0,r)} \,.$$

**Proof.** (a) By Proposition 5.21 we take

$$g(r) = \frac{1}{r^d L_{n-1}(1/r^2)l_n(1/r^2)^2}\,.$$

Recall that the functions $l_n$, respectively $L_n$, were defined in (5.36), respectively (5.37). Integration gives that

$$\int_0^r \rho^{d-1}g(\rho)\,d\rho = \int_0^r \frac{1}{\rho L_{n-1}(1/\rho^2)l_n(1/\rho^2)^2}\,d\rho = \frac{2}{l_n(1/r^2)}\,.$$

Therefore,

$$\kappa(r) = \frac{1}{L_n(1/r^2)} \geq \tilde{c} r^\gamma \,,$$

and the claim follows from Remark 5.59.
(b) This was shown in Example 5.57.                                      □

**Remark 5.62.** *We note that part (b) of both Propositions 5.60 and 5.61 are true for every pure jump process. This was proved in [131], and later also in [137].*

In the remainder of this subsection we discuss briefly the Krylov-Safonov type estimate involving the Lebesgue measure instead of the capacity. This type of estimate turns out to be very useful in case of a pure jump Lévy process. The method of proof comes from [13], while our exposition follows [145].

Assume that $Y = (Y_t : t \geq 0)$ is a subordinate Brownian motion via a subordinator with no drift. We retain the notation $j(|x|) = J(x)$, introduce functions

$$\eta_1(r) = r^{-2} \int_0^r \rho^{d+1} j(\rho)\, d\rho \,, \quad \eta_2(r) = \int_r^\infty \rho^{d-1} j(\rho)\, d\rho \,,$$

and let $\eta(r) = \eta_1(r) + \eta_2(r)$. The proof of the following result can be found in [145].

**Lemma 5.63.** *There exists a constant $C_{19} > 0$ such that for every $r \in (0,1)$, every $A \subset B(0,r)$ and any $y \in B(0,2r)$,*

$$\mathbb{P}^y(T_A < \tau_{B(0,3r)}) \geq C_{19} \frac{r^d j(4r)}{\eta(r)} \frac{|A|}{|B(0,r)|} \,,$$

*where $|\cdot|$ denotes the Lebesgue measure.*

**Proposition 5.64.** *Assume that $Y_t = X(S_t)$ where $S = (S_t : t \geq 0)$ is a pure jump subordinator, and the jumping function $J(x) = j(|x|)$ of $Y$ is such that $j$ satisfies $j(r) \sim r^{-d-\alpha}\ell(r)$, $r \to 0+$, with $0 < \alpha < 2$ and $\ell$ slowly varying at 0. Then there exists a constant $C_{20} > 0$ such that for every $r \in (0,1)$, every $A \subset B(0,r)$ and any $y \in B(0,2r)$,*

$$\mathbb{P}^y(T_A < \tau_{B(0,3r)}) \geq C_{20} \frac{|A|}{|B(0,r)|} \,.$$

**Proof.** It suffices to prove that $r^d j(4r)/\eta(r)$ is bounded from below by a positive constant. This is accomplished along the lines of the proof of Proposition 5.60.                                                         □

Note that the assumptions of Proposition 5.64 are satisfied for subordinate Brownian motions via $\alpha/2$-stable subordinators, relativistic $\alpha$-stable subordinators and the subordinators corresponding to Examples 5.15 and 5.16 (see Theorems 5.43 and 5.44).

In the case of, say, a geometric stable process $Y$, one obtains from Lemma 5.63 a weak form of the hitting time estimate: There exists $C_{21} > 0$ such that for every $r \in (0, 1/2)$, every $A \subset B(0, r)$ and any $y \in B(0, 2r)$,

$$\mathbb{P}^y(T_A < \tau_{B(0,3r)}) \geq C_{21} \frac{1}{\log(1/r)} \frac{|A|}{|B(0,r)|}. \tag{5.95}$$

## 5.4.3  Proof of Harnack Inequality

Let $Y = (Y_t : t \geq 0)$ be a subordinate Brownian motion in $\mathbb{R}^d$ and let $D$ be an open subset of $\mathbb{R}^d$. A function $h : \mathbb{R}^d \to [0, +\infty]$ is said to be harmonic in $D$ with respect to the process $Y$ if for every bounded open set $B \subset \overline{B} \subset D$,

$$h(x) = \mathbb{E}^x[h(Y_{\tau_B})], \quad \forall x \in B,$$

where $\tau_B = \inf\{t > 0 : Y_t \notin B\}$ is the exit time of $Y$ from $B$. Harnack inequality is a statement about the growth rate of nonnegative harmonic functions in compact subsets of $D$. We will first discuss two proofs of a scale invariant Harnack inequality for small balls. Next, we will give a proof of a weak form of Harnack inequality for small balls for the iterated geometric stable process. All discussed forms of the inequality lead to the following Harnack inequality: For any compact set $K \subset D$, there exists a constant $C > 0$, depending only on $D$ and $K$, such that for every nonnegative harmonic function $h$ with respect to $Y$ in $D$, it holds that

$$\sup_{x \in K} h(x) \leq C \inf_{x \in K} h(x).$$

The general methodology of proving Harnack inequality for jump processes is explained in [145] following the pioneering work [13] (for an alternative approach see [39]). The same method was also used in [14] and [49] to prove a parabolic Harnack inequality. There are two essential ingredients: The first one is a Krylov-Safonov-type estimate for the hitting probability discussed in the previous subsection. The form given in Lemma 5.63 and Proposition 5.64 can be used in the case of pure jump processes for which one has good control of the behavior of the jumping function $J$ at zero. More precisely, one needs that $j(r)$ is a regularly varying function of index $-d - \alpha$ for $0 < \alpha < 2$ when $r \to 0+$. This, as shown in Proposition 5.64, implies that the function of $r$ on the right-hand side of the estimate can be replaced by a constant, which is desirable to obtaining the scale invariant form of Harnack inequality for small balls. In the case of a geometric stable process the behavior of $J$ near zero is known (see Theorem 5.45), but leads to the inequality (5.95) having the factor $1/\log(1/r)$ on the right-hand side. This yields a weak type of Harnack inequality for balls. In the case of the iterated geometric stable processes, no information about the behavior of $J$ near zero is available, and hence one does not have any control on the factor $r^d j(r)/\eta(r)$ in Lemma 5.63. In the case where $Y$ has a continuous component (i.e, the subordinator $S$ has a drift), or the case when information on the behavior of $J$ near zero is missing, one can use the form of Krylov-Safonov inequality described in Propositions 5.60 and 5.61.

The second ingredient in the proof is the following result which can be considered as a very weak form of Harnack inequality (more precisely, Harnack inequality for harmonic measures of sets away from the ball). Recall that $R > 0$ was defined in (5.89).

**Proposition 5.65.** *Let $Y$ be a subordinate Brownian motion such that the function $j$ defined in (5.62) satisfies conditions (5.65) and (5.66). There exists a positive constant $C_{22} > 0$ such that for any $r \in (0, R)$, any $y, z \in B(0, r/2)$ and any nonnegative function $H$ supported on $B(0, 2r)^c$ it holds that*

$$\mathbb{E}^z H(Y(\tau_{B(0,r)})) \leq C_{22} \mathbb{E}^y H(Y(\tau_{B(0,r)})). \tag{5.96}$$

**Proof.** This is an immediate consequence of Proposition 5.50 and the comparison results for the mean exit times explained in Remark 5.62 (see also Propositions 5.60 and 5.61). □

We are now ready to state Harnack inequality under two different set of conditions.

**Theorem 5.66.** *Let $Y$ be a subordinate Brownian motion such that the function $j$ defined in (5.62) satisfies conditions (5.65) and (5.66) and is further regularly varying at zero with index $-d - \alpha$ where $0 < \alpha < 2$. Then there exists a constant $C > 0$ such that, for any $r \in (0, 1/4)$, and any function $h$ which is nonnegative, bounded on $\mathbb{R}^d$, and harmonic with respect to $Y$ in $B(0, 16r)$, we have*

$$h(x) \leq C h(y), \quad \forall x, y \in B(0, r).$$

Proof of this Harnack inequality follows from [145] and uses Proposition 5.64. The second set of conditions for Harnack inequality uses Proposition 5.60. Recall the constant $c$ defined in (5.87).

**Theorem 5.67.** *Let $Y$ be a transient subordinate Brownian motion such that the function $j$ defined in (5.62) satisfies conditions (5.65) and (5.66), and assume further that the subordinator $S$ is special and its Laplace exponent $\phi$ satisfies $\phi(\lambda) \sim b\lambda^{\alpha/2}$, $\lambda \to \infty$, with $\alpha \in (0, 2]$ and $b > 0$. Then there exists a constant $C > 0$ such that, for any $r \in (0, (14c)^{-1}R)$, and any function $h$ which is nonnegative, bounded on $\mathbb{R}^d$, and harmonic with respect to $Y$ in $B(0, 14cr)$, we have*

$$h(x) \leq C h(y), \quad \forall x, y \in B(0, r/2).$$

Under these conditions, Harnack inequality was proved in [132]. Unfortunately, despite the fact that Proposition 5.60 holds under weaker conditions for $\phi$ than the ones stated in the theorem above, we were unable to carry out a proof in this more general case.

Now we are going to present a proof of a weak form of Harnack inequality for iterated geometric stable processes. Let $S^{(n)}$ be the iterated geometric stable subordinator and let $Y_t^{(n)} = X(S_t^{(n)})$ be the corresponding subordinate process. We assume that $d > 2(\alpha/2)^n$. For simplicity we write $Y$ instead of $Y^{(n)}$. We state again Propositions 5.61 (a) and 5.65:

Let $\gamma > 0$. There exists a constant $C_{17} > 0$ such that for any $r \in (0, (7c)^{-1}R)$, any closed subset $A$ of $B(0,r)$, and any $y \in B(0,r)$

$$\mathbb{P}^y(T_A < \tau_{B(0,7cr)}) \geq C_{17}\, r^\gamma \frac{\text{Cap}(A)}{\text{Cap}(\overline{B(0,r)})}, \tag{5.97}$$

There exists a positive constant $C_{22} > 0$ such that for any $r \in (0, R)$, any $y, z \in B(0, r/2)$ and any nonnegative function $H$ supported on $B(0,r)^c$ it holds that

$$\mathbb{E}^z H(Y(\tau_{B(0,r)})) \leq C_{22}\mathbb{E}^y H(Y(\tau_{B(0,r)})). \tag{5.98}$$

We will also need the following lemma.

**Lemma 5.68.** *There exists a positive constant $C_{23}$ such that for all $0 < \rho < r < 1/e_{n+1}$,*

$$\frac{\text{Cap}(\overline{B(0,\rho)})}{\text{Cap}(\overline{B(0,r)})} \geq C_{23} \left(\frac{\rho}{r}\right)^d.$$

**Proof.** By Example 5.57,

$$C_5 r^d l_n(1/r) \leq \text{Cap}(\overline{B(0,r)}) \leq C_6 r^d l_n(1/r)$$

for every $r < 1/e_{n+1}$. Therefore,

$$\frac{\text{Cap}(\overline{B(0,\rho)})}{\text{Cap}(\overline{B(0,r)})} \geq \frac{C_5 \rho^d l_n(1/\rho)}{C_6 r^d l_n(1/r)} \geq \frac{C_5}{C_6} \left(\frac{\rho}{r}\right)^d,$$

where the last inequality follows from the fact that $l_n$ is increasing at infinity.
□

**Theorem 5.69.** *Let $R$ and $c$ be defined by (5.89) and (5.87) respectively. Let $r \in (0, (14c)^{-1}R)$. There exists a constant $C > 0$ such that for every nonnegative bounded function $h$ in $\mathbb{R}^d$ which is harmonic with respect to $Y$ in $B(0, 14cr)$ it holds*

$$h(x) \leq Ch(y), \quad x, y \in B(0, r/2).$$

**Remark 5.70.** *Note that the constant $C$ in the theorem may depend on the radius $r$. This is why the above Harnack inequality is weak. A version of a weak Harnack inequality appeared in [12], and our proof follows the arguments there. A similar proof, in a somewhat different context, was given in [147].*

**Proof.** We fix $\gamma \in (0,1)$. Suppose that $h$ is nonnegative and bounded in $\mathbb{R}^d$ and harmonic with respect to $Y$ in $B(0, 14cr)$. By looking at $h + \epsilon$ and letting $\epsilon \downarrow 0$, we may suppose that $h$ is bounded from below by a positive constant. By looking at $ah$ for a suitable $a > 0$, we may suppose that $\inf_{B(0,r/2)} h = 1/2$. We want to bound $h$ from above in $B(0, r/2)$ by a constant depending only on $r$, $d$ and $\gamma$. Choose $z_1 \in B(0, r/2)$ such that $h(z_1) \le 1$. Choose $\rho \in (1, \gamma^{-1})$. For $i \ge 1$ let

$$r_i = \frac{c_1 r}{i^\rho},$$

where $c_1$ is a constant to be determined later. We require first of all that $c_1$ is small enough so that

$$\sum_{i=1}^{\infty} r_i \le \frac{r}{8}. \tag{5.99}$$

Recall that there exists $c_2 := C_{17} > 0$ such that for any $s \in (0, (7c)^{-1}R)$, any closed subset $A \subset B(0, s)$ and any $y \in B(0, s)$,

$$\mathbb{P}^y(T_A < \tau_{B(0,7cs)}) \ge c_2 s^\gamma \frac{\mathrm{Cap}(A)}{\mathrm{Cap}(\overline{B(0, s)})}. \tag{5.100}$$

Let $c_3$ be a constant such that

$$c_3 \le c_2 2^{-4-\gamma+\rho\gamma}.$$

Denote the constant $C_8$ from Lemma 5.68 by $c_4$. Once $c_1$ and $c_3$ have been chosen, choose $K_1$ sufficiently large so that

$$\frac{1}{4}(7c)^{-d-\gamma} c_2 c_4 K_1 \exp((14c)^{-\gamma} r^\gamma c_1 c_3 i^{1-\rho\gamma}) c_1^{4\gamma+d} r^{4\gamma} \ge 2 i^{4\rho\gamma+\rho d} \tag{5.101}$$

for all $i \ge 1$. Such a choice is possible since $\rho\gamma < 1$. Note that $K_1$ will depend on $r, d$ and $\gamma$ as well as constants $c, c_1, c_2, c_3$ and $c_4$. Suppose now that there exists $x_1 \in B(0, r/2)$ with $h(x_1) \ge K_1$. We will show that in this case there exists a sequence $\{(x_j, K_j) : j \ge 1\}$ with $x_{j+1} \in B(x_j, 2r_j) \subset B(0, 3r/4)$, $K_j = h(x_j)$, and

$$K_j \ge K_1 \exp((14c)^{-\gamma} r^\gamma c_1 c_3 j^{1-\rho\gamma}). \tag{5.102}$$

Since $1 - \rho\gamma > 0$, we have $K_j \to \infty$, a contradiction to the assumption that $h$ is bounded. We can then conclude that $h$ must be bounded by $K_1$ on $B(0, r/2)$, and hence $h(x) \le 2K_1 h(y)$ if $x, y \in B(0, r/2)$.

Suppose that $x_1, x_2, \ldots, x_i$ have been selected and that (5.102) holds for $j = 1, \ldots, i$. We will show that there exists $x_{i+1} \in B(x_i, 2r_i)$ such that if $K_{i+1} = h(x_{i+1})$, then (5.102) holds for $j = i + 1$; we then use induction to conclude that (5.102) holds for all $j$.

Let
$$A_i = \{y \in B(x_i, (14c)^{-1} r_i) : h(y) \geq K_i r_i^{2\gamma}\}.$$

First we prove that
$$\frac{\mathrm{Cap}(A_i)}{\mathrm{Cap}(B(x_i, (14c)^{-1} r_i))} \leq \frac{1}{4}. \tag{5.103}$$

To prove this claim, we suppose to the contrary that $\mathrm{Cap}(A_i)/\mathrm{Cap}(B(x_i, (14c)^{-1} r_i)) > 1/4$. Let $F$ be a compact subset of $A_i$ with $\mathrm{Cap}(F)/\mathrm{Cap}(B(x_i, (14c)^{-1} r_i)) > 1/4$. Recall that $r \geq 8r_i$. Now we have

$$
\begin{aligned}
1 \geq h(z_1) &\geq \mathbb{E}^{z_1}[h(Y_{T_F \wedge \tau_{B(0,7cr)}}); T_F < \tau_{B(0,7cr)}] \\
&\geq K_i r_i^{2\gamma} \mathbb{P}^{z_1}(T_F < \tau_{B(0,7cr)}) \\
&\geq c_2 K_i r_i^{2\gamma} r^\gamma \frac{\mathrm{Cap}(F)}{\mathrm{Cap}(B(0,r))} \\
&= c_2 K_i r_i^{2\gamma} r^\gamma \frac{\mathrm{Cap}(F)}{\mathrm{Cap}(B(x_i, (7c)^{-1} r_i))} \frac{\mathrm{Cap}(B(x_i, (7c)^{-1} r_i))}{\mathrm{Cap}(B(0,r))} \\
&\geq \frac{1}{4} c_2 K_i r_i^{2\gamma} r^\gamma \frac{\mathrm{Cap} B(0, (7c)^{-1} r_i))}{\mathrm{Cap}(B(0,r))} \\
&\geq \frac{1}{4} c_2 K_i r_i^{2\gamma} r^\gamma c_4 \left(\frac{(7c)^{-1} r_i}{r}\right)^d \\
&= \frac{1}{4} c_2 c_4 (7c)^{-d} K_i r_i^{2\gamma} r^\gamma \left(\frac{r_i}{r}\right)^d \\
&\geq \frac{1}{4} c_2 c_4 (7c)^{-d-\gamma} K_i r_i^{4\gamma} \left(\frac{r_i}{r}\right)^d \\
&\geq \frac{1}{4} c_2 c_4 (7c)^{-d-\gamma} K_1 \exp((14c)^{-\gamma} r^\gamma c_1 c_3 i^{1-\rho\gamma}) r_i^{4\gamma} \left(\frac{r_i}{r}\right)^d \\
&\geq \frac{1}{4} c_2 c_4 (7c)^{-d-\gamma} K_1 \exp((14c)^{-\gamma} r^\gamma c_1 c_3 i^{1-\rho\gamma}) \left(\frac{c_1 r}{i^\rho}\right)^{4\gamma} \left(\frac{c_1}{i^\rho}\right)^d \\
&\geq \frac{1}{4} c_2 c_4 (7c)^{-d-\gamma} K_1 \exp((14c)^{-\gamma} r^\gamma c_1 c_3 j^{1-\rho\gamma}) c_1^{4\gamma+d} r^{4\gamma} i^{-4\gamma\rho-\rho d} \\
&\geq 2 i^{4\gamma\rho+\rho d} i^{-4\gamma\rho-\rho d} = 2.
\end{aligned}
$$

We used the definition of harmonicity in the first line, (5.100) in the third, Lemma 5.68 in the sixth, (5.102) in the ninth, and (5.101) in the last line. This is a contradiction, and therefore (5.103) is valid.

By subadditivity of the capacity and by (5.103) it follows that there exists $E_i \subset B(x_i, (14c)^{-1} c) \setminus A_i$ such that

$$\frac{\mathrm{Cap}(E_i)}{\mathrm{Cap}(B(x_i, (14c)^{-1} r_i))} \geq \frac{1}{2}.$$

Write $\tau_i$ for $\tau_{B(x_i, r_i/2)}$ and let $p_i := \mathbb{P}^{x_i}(T_{E_i} < \tau_i)$. It follows from (5.100) that

$$
\begin{aligned}
p_i &\geq c_2 \left(\frac{r_i}{14c}\right)^\gamma \frac{\operatorname{Cap}(E_i)}{\operatorname{Cap}(B(x_i, (14c)^{-1}))} \\
&\geq \frac{c_2}{2} \left(\frac{r_i}{14c}\right)^\gamma .
\end{aligned}
\tag{5.104}
$$

Set $M_i = \sup_{B(x_i, r_i)} h$. Then

$$
\begin{aligned}
K_i = h(x_i) &= \mathbb{E}^{x_i}[h(Y_{T_{E_i} \wedge \tau_i}); T_{E_i} < \tau_i] \\
&+ \mathbb{E}^{x_i}[h(Y_{T_{E_i} \wedge \tau_i}); T_{E_i} \geq \tau_i, Y_{\tau_i} \in B(x_i, r_i)] \\
&+ \mathbb{E}^{x_i}[h(Y_{T_{E_i} \wedge \tau_i}); T_{E_i} \geq \tau_i, Y_{\tau_i} \notin B(x_i, r_i)].
\end{aligned}
\tag{5.105}
$$

We are going to estimate each term separately. Since $E_i$ is compact, we have

$$
\mathbb{E}^{x_i}[h(Y_{T_{E_i} \wedge \tau_i}); T_{E_i} < \tau_i] \leq K_i r_i^{2\gamma} \mathbb{P}^{x_i}(T_{E_i} < \tau_i) \leq K_i r_i^{2\gamma}.
$$

Further,

$$
\mathbb{E}^{x_i}[h(Y_{T_{E_i} \wedge \tau_i}); T_{E_i} \geq \tau_i, Y_{\tau_i} \in B(x_i, r_i)] \leq M_i(1 - p_i).
$$

Inequality (5.103) implies in particular that there exists $y_i \in B(x_i, (14c)^{-1} r_i)$ with $h(y_i) \leq K_i r_i^{2\gamma}$. We then have, by (5.98) and with $c_5 = C_{22}$

$$
\begin{aligned}
K_i r_i^{2\gamma} \geq h(y_i) &\geq \mathbb{E}^{y_i}[h(Y_{\tau_i}) : Y_{\tau_i} \notin B(x_i, r_i)] \\
&\geq c_5 \mathbb{E}^{x_i}[h(Y_{\tau_i}) : Y_{\tau_i} \notin B(x_i, r_i)].
\end{aligned}
\tag{5.106}
$$

Therefore

$$
\mathbb{E}^{x_i}[h(Y_{T_{E_i} \wedge \tau_i}); T_{E_i} \geq \tau_i, Y_{\tau_i} \notin B(x_i, r_i)] \leq c_6 K_i r_i^{2\gamma}
$$

for the positive constant $c_6 = 1/c_5$. Consequently we have

$$
K_i \leq (1 + c_6) K_i r_i^{2\gamma} + M_i(1 - p_i).
\tag{5.107}
$$

Rearranging, we get

$$
M_i \geq K_i \left(\frac{1 - (1 + c_6) r_i^{2\gamma}}{1 - p_i}\right).
\tag{5.108}
$$

Now choose

$$
c_1 \leq \min\{\frac{1}{14c} \frac{1}{r} \left(\frac{1}{4} \frac{c_2}{1 + c_6}\right)^{1/\gamma}, 1\}.
$$

This choice of $c_1$ implies that

$$2(1+c_6)r_i^{2\gamma} \leq \frac{c_2}{2}\left(\frac{r_i}{14c}\right)^\gamma \leq p_i,$$

where the second inequality follows from (5.104). Therefore, $1-(1+c_6)r_i^{2\gamma} \geq 1 - p_i/2$, and hence by use of (5.108)

$$M_i \geq K_i\left(\frac{1-\frac{1}{2}p_i}{1-p_i}\right) > (1+\frac{p_i}{2})K_i.$$

Using the definition of $M_i$ and (5.98), there exists a point $x_{i+1} \in \overline{B(x_i,r_i)} \subset B(x_i,2r_i)$ such that

$$K_{i+1} = h(x_{i+1}) \geq K_i\left(1 + \frac{c_2}{4}\left(\frac{r_i}{14c}\right)^\gamma\right).$$

Taking logarithms and writing

$$\log K_{i+1} = \log K_i + \sum_{j=1}^{i}[\log K_{j+1} - \log K_j],$$

we have

$$\log K_{i+1} \geq \log K_1 + \sum_{j=1}^{i}\log\left(1 + \frac{c_2}{4}\left(\frac{r_j}{14c}\right)^\gamma\right)$$

$$\geq \log K_1 + \sum_{j=1}^{i}\frac{c_2}{4}\frac{r_j^\gamma}{(14c)^\gamma}$$

$$= \log K_1 + \frac{c_2}{4}\frac{1}{(14c)^\gamma}\sum_{j=1}^{i}\left(\frac{c_1 r}{j^\rho}\right)^\gamma$$

$$\geq \log K_1 + \frac{c_2}{4}\frac{1}{(14c)^\gamma}r^\gamma c_1^\gamma\sum_{j=1}^{i}j^{-\rho\gamma}$$

$$\geq \log K_1 + \frac{c_2}{4}\frac{1}{(14c)^\gamma}r^\gamma c_1 i^{1-\rho\gamma}$$

$$\geq \log K_1 + \frac{1}{(14c)^\gamma}r^\gamma c_1 c_3 (i+1)^{1-\rho\gamma}.$$

In the fifth line we used the fact that $c_1 < 1$. For the last line recall that

$$c_3 \leq c_2 2^{-4-\gamma+\rho\gamma} = \frac{c_2}{2^{3+\gamma}}\left(\frac{1}{2}\right)^{1-\rho\gamma} \leq \frac{c_2}{2^{3+\gamma}}\left(\frac{i}{i+1}\right)^{1-\rho\gamma},$$

implying that

$$\frac{c_2}{4} i^{1-\rho\gamma} \geq 2^{1+\gamma} c_3 (1+i)^{1-\rho\gamma}.$$

Therefore we have obtained that

$$K_{i+1} \geq K_1 \exp((14c)^{-\gamma} r^\gamma c_1 c_3 (i+1)^{1-\rho\gamma})$$

which is (5.102) for $i+1$. The proof is now finished.                    □

**Remark 5.71.** *The proof given above can be easily modified to provide a proof of Theorem 5.67. Indeed, one can modify slightly Lemma 5.68, take $\gamma = 0$ and choose any $\rho > 1$ in the proof. The choice of $K_1$ in (5.101) and $K_j$ in (5.102) will not depend on $r > 0$, thus giving a strong form of Harnack inequality.*

## 5.5  Subordinate Killed Brownian Motion

### 5.5.1  Definitions

Let $X = (X_t, \mathbb{P}^x)$ be a $d$-dimensional Brownian motion. Let $D$ be a bounded connected open set in $\mathbb{R}^d$, and let $\tau_D = \inf\{t > 0 : X_t \notin D\}$ be the exit time of $X$ from $D$. Define

$$X_t^D = \begin{cases} X_t, & t < \tau_D, \\ \partial, & t \geq \tau_D, \end{cases}$$

where $\partial$ is the cemetery. We call $X^D$ a Brownian motion killed upon exiting $D$, or simply, a killed Brownian motion. The semigroup of $X^D$ will be denoted by $(P_t^D : t \geq 0)$, and its transition density by $p^D(t, x, y)$, $t > 0$, $x, y \in D$. The transition density $p^D(t, x, y)$ is strictly positive, and hence the eigenfunction $\varphi_0$ of the operator $-\Delta|_D$ corresponding to the smallest eigenvalue $\lambda_0$ can be chosen to be strictly positive, see, for instance, [66]. The potential operator of $X^D$ is given by

$$G^D f(x) = \int_0^\infty P_t^D f(x)\, dt$$

and has a density $G^D(x, y)$, $x, y \in D$. Here, and further below, $f$ denotes a nonnegative Borel function on $D$. We recall the following well-known facts: If $h$ is a nonnegative harmonic function for $X^D$ (i.e., harmonic for $\Delta$ in $D$), then both $h$ and $P_t^D h$ are continuous functions in $D$.

In this section we always assume that $(P_t^D : t \geq 0)$ is *intrinsically ultracontractive*, that is, for each $t > 0$ there exists a constant $c_t$ such that

$$p^D(t, x, y) \leq c_t \varphi_0(x)\varphi_0(y), \quad x, y \in D, \tag{5.109}$$

where $\varphi_0$ is the positive eigenfunction corresponding to the smallest eigen-value $\lambda_0$ of the Dirichlet Laplacian $-\Delta|_D$. It is well known that (see, for instance, [67]) when $(P_t^D ; t \geq 0)$ is intrinsically ultracontractive there is $\tilde{c}_t > 0$ such that

$$p^D(t,x,y) \geq \tilde{c}_t \varphi_0(x)\varphi_0(y), \quad x,y \in D.$$

Intrinsic ultracontractivity was introduced by Davies and Simon in [67]. It is well known that (see, for instance, [6]) $(P_t^D : t \geq 0)$ is intrinsically ultracontractive when $D$ is a bounded Lipschitz domain, or a Hölder domain of order 0, or a uniformly Hölder domain of order $\beta \in (0,2)$.

Let $S = (S_t : t \geq 0)$ and $T = (T_t : t \geq 0)$ be two special subordinators. Suppose that $X$, $S$ and $T$ are independent. We assume that the Laplace exponents of $S$ and $T$, denoted by $\phi$ and $\psi$ respectively, are conjugate, i.e., $\lambda = \phi(\lambda)\psi(\lambda)$. We also assume that $\phi$ has the representation (5.2) with $b > 0$ or $\mu(0,\infty) = \infty$. We define two subordinate processes $Y^D$ and $Z^D$ by

$$Y_t^D = X^D(S_t), \quad t \geq 0$$
$$Z_t^D = X^D(T_t), \quad t \geq 0.$$

Then $Y^D = (Y_t^D : t \geq 0)$ and $Z^D = (Z_t^D : t \geq 0)$ are strong Markov processes on $D$. We call $Y^D$ (resp. $Z^D$) a subordinate killed Brownian motion. If we use $\eta_t(ds)$ and $\theta_t(ds)$ to denote the distributions of $S_t$ and $T_t$ respectively, the semigroups of $Y^D$ and $Z^D$ are given by

$$Q_t^D f(x) = \int_0^\infty P_s^D f(x)\eta_t(ds),$$
$$R_t^D f(x) = \int_0^\infty P_s^D f(x)\theta_t(ds),$$

respectively. The semigroup $Q_t^D$ has a density given by

$$q^D(t,x,y) = \int_0^\infty p^D(s,x,y)\eta_t(ds).$$

The semigroup $R_t^D$ will have a density

$$r^D(t,x,y) = \int_0^\infty p^D(s,x,y)\theta_t(ds)$$

in the case $b = 0$, while for $b > 0$, $R_t^D$ is not absolutely continuous with respect to the Lebesgue measure. Let $U$ and $V$ denote the potential measures of $S$ and $T$, respectively. Then there are decreasing functions $u$ and $v$ defined on $(0,\infty)$ such that $U(dt) = u(t)\,dt$ and $V(dt) = b\epsilon_0(dt) + v(t)\,dt$. The potential kernels of $Y^D$ and $Z^D$ are given by

$$U^D f(x) = \int_0^\infty P_t^D f(x) \, U(dt) = \int_0^\infty P_t^D f(x) \, u(t) \, dt,$$

$$V^D f(x) = \int_0^\infty P_t^D f(x) \, V(dt) = bf(x) + \int_0^\infty P_t^D f(x) \, v(t) \, dt,$$

respectively. The potential kernel $U^D$ has a density given by

$$U^D(x, y) = \int_0^\infty p^D(t, x, y) \, u(t) \, dt,$$

while $V^D$ needs not be absolutely continuous with respect to the Lebesgue measure. Note that $U^D(x, y)$ is the Green function of the process $Y^D$. For the process $Y^D$ we define the potential of a Borel measure $m$ on $D$ by

$$U^D m(x) := \int_D U^D(x, y) \, m(dy) = \int_0^\infty P_t^D m(x) \, u(t) \, dt.$$

Let $(U_\lambda^D, \lambda > 0)$ be the resolvent of the semigroup $(Q_t^D, t \geq 0)$. Then $U_\lambda^D$ is given by a kernel which is absolutely continuous with respect to the Lebesgue measure. Moreover, one can easily show that for a bounded Borel function $f$ vanishing outside a compact subset of $D$, the functions $x \mapsto U_\lambda^D f(x)$, $\lambda > 0$, and $x \mapsto U^D f(x)$ are continuous. This implies (e.g., [23], p.266) that excessive functions of $Y^D$ are lower semicontinuous.

Recall that a measurable function $s : D \to [0, \infty]$ is excessive for $Y^D$ (or $Q_t^D$), if $Q_t^D s \leq s$ for all $t \geq 0$ and $s = \lim_{t \to 0} Q_t^D s$. We will denote the family of all excessive function for $Y^D$ by $\mathcal{S}(Y^D)$. The notations $\mathcal{S}(X^D)$ and $\mathcal{S}(Z^D)$ are now self-explanatory.

A measurable function $h : D \to [0, \infty]$ is harmonic for $Y^D$ if $h$ is not identically infinite in $D$ and if for every relatively compact open subset $U \subset \overline{U} \subset D$,

$$h(x) = \mathbb{E}^x [h(Y^D(\tau_U^Y))], \quad \forall x \in U,$$

where $\tau_U^Y = \inf\{t : Y_t^D \notin U\}$ is the first exit time of $Y^D$ from $U$. We will denote the family of all nonnegative harmonic function for $Y^D$ by $\mathcal{H}^+(Y^D)$. Similarly, $\mathcal{H}^+(X^D)$ will denote the family of all nonnegative harmonic functions for $X^D$. It is well known that $\mathcal{H}^+(\cdot) \subset \mathcal{S}(\cdot)$.

### 5.5.2  Representation of Excessive and Harmonic Functions of Subordinate Process

The factorization in the next proposition is similar in spirit to Theorem 4.1 (5) in [136].

**Proposition 5.72.** *(a)  For any nonnegative Borel function f on D we have*

$$U^D V^D f(x) = V^D U^D f(x) = G^D f(x), \quad x \in D.$$

*(b)  For any Borel measure m on D we have*

$$V^D U^D m(x) = G^D m(x).$$

**Proof.** (a) We are only going to show that $U^D V^D f(x) = G^D f(x)$ for all $x \in D$. For the proof of $V^D U^D f(x) = G^D f(x)$ see part (b). For any nonnegative Borel function $f$ on $D$, by using the Markov property and Theorem 5.6 we get that

$$
\begin{aligned}
U^D V^D f(x) &= \int_0^\infty P_t^D V^D f(x) u(t) dt \\
&= \int_0^\infty P_t^D \left( bf(x) + \int_0^\infty P_s^D f(x) v(s) ds \right) u(t) dt \\
&= b U^D f(x) + \int_0^\infty P_t^D \left( \int_0^\infty P_s^D f(x) v(s) ds \right) u(t) dt \\
&= b U^D f(x) + \int_0^\infty \int_0^\infty P_{t+s}^D f(x) v(s) ds\, u(t) dt \\
&= b U^D f(x) + \int_0^\infty \int_t^\infty P_r^D f(x) v(r-t) dr\, u(t) dt \\
&= b U^D f(x) + \int_0^\infty \left( \int_0^r u(t) v(r-t) dt \right) P_r^D f(x) dr \\
&= \int_0^\infty \left( bu(r) + \int_0^r u(t) v(r-t) dt \right) P_r^D f(x) dr \\
&= \int_0^\infty P_r^D f(x) dr = G^D f(x).
\end{aligned}
$$

(b) Similarly as above,

$$
\begin{aligned}
V^D U^D m(x) &= b U^D m(x) + \int_0^\infty P_t^D U^D m(x) v(t)\, dt \\
&= b U^D m(x) + \int_0^\infty P_t^D \left( \int_0^\infty P_s^D m(x) u(s)\, ds \right) v(t)\, dt \\
&= b U^D m(x) + \int_0^\infty \int_0^\infty P_{t+s}^D m(x) u(s)\, ds\, v(t)\, dt \\
&= b U^D m(x) + \int_0^\infty \int_t^\infty P_r^D m(x) u(r-t)\, dr\, v(t)\, dt \\
&= b U^D m(x) + \int_0^\infty \left( \int_0^r u(r-t) v(t)\, dt \right) P_r^D m(x)\, dr
\end{aligned}
$$

$$= \int_0^\infty \left( b + \int_0^r u(r-t)v(t)\, dt \right) P_r^D m(x)\, dr$$

$$= \int_0^\infty P_r^D m(x)\, dr = G^D m(x).$$

□

**Proposition 5.73.** *Let $g$ be an excessive function for $Y^D$. Then $V^D g$ is excessive for $X^D$.*

**Proof.** We first observe that if $g$ is excessive with respect to $Y^D$, then $g$ is the increasing limit of $U^D f_n$ for some $f_n$. Hence it follows from Proposition 5.72 that

$$V^D g = \lim_{n\to\infty} V^D U^D f_n = \lim_{n\to\infty} G^D f_n,$$

which implies that $V^D g$ is either identically infinite or excessive with respect to $X^D$. We prove now that $V^D g$ is not identically infinite. In fact, since $g$ is excessive with respect to $Y^D$, there exists $x_0 \in D$ such that for every $t > 0$,

$$\infty > g(x_0) \geq Q_t^D g(x_0) = \int_0^\infty P_s^D g(x_0)\rho_t(ds).$$

Thus there is $s > 0$ such that $P_s^D g(x_0)$ is finite. Hence

$$\infty > P_s^D g(x_0) = \int_D p^D(s, x_0, y)g(y)\, dy \geq \tilde{c}_s \varphi_0(x_0) \int_D \varphi_0(y)g(y)\, dy,$$

so we have $\int_D \varphi_0(y)g(y)\, dy < \infty$. Consequently

$$\int_D V^D g(x)\varphi_0(x)\, dx = \int_D g(x) V^D \varphi_0(x)\, dx$$

$$= \int_D g(x) \left( b\varphi_0(x) + \int_0^\infty P_t^D \varphi_0(x)v(t)\, dt \right) dx$$

$$= \int_D g(x) \left( b\varphi_0(x) + \int_0^\infty e^{-\lambda_0 t}\varphi_0(x)v(t)\, dt \right) dx$$

$$= \int_D \varphi_0(x)g(x)\, dx \left( b + \int_0^\infty e^{-\lambda_0 t}v(t)\, dt \right) < \infty.$$

Therefore $s = V^D g$ is not identically infinite in $D$.

□

**Remark 5.74.** *Note that the proposition above is valid with $Y^D$ and $Z^D$ interchanged: If $g$ is excessive for $Z^D$, then $U^D g$ is excessive for $X^D$. Using this we can easily get the following simple fact: If $f$ and $g$ are two nonnegative Borel functions on $D$ such that $V^D f$ and $V^D g$ are not identically infinite, and such that $V^D f = V^D g$ a.e., then $f = g$ a.e. In fact, since $V^D f$ and $V^D g$ are excessive for $Z^D$, we know that $G^D f = U^D V^D f$ and $G^D g = U^D V^D g$*

are excessive for $X^D$. Moreover, by the absolute continuity of $U^D$, we have that $G^D f = G^D g$. The a.e. equality of $f$ and $g$ follows from the uniqueness principle for $G^D$.

The second part of Proposition 5.72 shows that if $s = G^D m$ is the potential of a measure, then $s = V^D g$ where $g = U^D m$ is excessive for $Y^D$. The function $g$ can be written in the following way:

$$
\begin{aligned}
g(x) &= \int_0^\infty P_s^D m(x) u(s)\, ds \\
&= \int_0^\infty P_s^D m(x) \left( u(\infty) + \int_s^\infty -du(t) \right) ds \\
&= \int_0^\infty P_s^D m(x) u(\infty)\, ds + \int_0^\infty P_s^D m(x) \left( \int_s^\infty -du(t) \right) ds \\
&= u(\infty) s(x) + \int_0^\infty \left( \int_0^t P_s^D m(x)\, ds \right) (-du(t)) \\
&= u(\infty) s(x) + \int_0^\infty (P_t^D s(x) - s(x))\, du(t)\,.
\end{aligned}
\tag{5.110}
$$

In the next proposition we will show that every excessive function $s$ for $X^D$ can be represented as a potential $V^D g$, where $g$, given by (5.110), is excessive for $Y^D$. We need the following important lemma.

**Lemma 5.75.** *Let $h$ be a nonnegative harmonic function for $X^D$, and let*

$$
g(x) = u(\infty) h(x) + \int_0^\infty (P_t^D h(x) - h(x))\, du(t)\,.
\tag{5.111}
$$

*Then $g$ is continuous.*

**Proof.** For any $\epsilon > 0$ it holds that $\left| \int_\epsilon^\infty du(t) \right| \le u(\epsilon)$. Hence from the continuity of $h$ and $P_t^D h$ it follows by the dominated convergence theorem that the function

$$
x \mapsto \int_\epsilon^\infty (P_t^D h(x) - h(x))\, du(t), \quad x \in D,
$$

is continuous. Therefore we only need to prove that the function

$$
x \mapsto \int_0^\epsilon (P_t^D h(x) - h(x))\, du(t), \quad x \in D,
$$

is continuous. For any $x_0 \in D$ choose $r > 0$ such that $B(x_0, 2r) \subset D$, and let $B = B(x_0, r)$. It is enough to show that

$$
\lim_{\epsilon \downarrow 0} \int_0^\epsilon (P_t^D h(x) - h(x))\, du(t) = 0
$$

uniformly on $\overline{B}$, the closure of $B$. For any $x \in B$, $h(X_{t \wedge \tau_B})$ is a $\mathbb{P}^x$-martingale. Therefore,

$$
\begin{aligned}
0 \leq h(x) - P_t^D h(x) &= \mathbb{E}^x[h(X_{t \wedge \tau_B})] - \mathbb{E}^x[h(X_t), t < \tau_D] \\
&= \mathbb{E}^x[h(X_t), t < \tau_B] + \mathbb{E}^x[h(X_{\tau_B}), \tau_B \leq t] \\
&\quad - \mathbb{E}^x[h(X_t), t < \tau_B] - \mathbb{E}^x[h(X_t), \tau_B \leq t < \tau_D] \\
&= \mathbb{E}^x[h(X_{\tau_B}), \tau_B \leq t] - \mathbb{E}^x[h(X_t), \tau_B \leq t < \tau_D] \\
&\leq \mathbb{E}^x[h(X_{\tau_B}), \tau_B \leq t] \leq M \mathbb{P}^x(\tau_B \leq t),
\end{aligned}
\tag{5.112}
$$

where $M$ is a constant such that $h(y) \leq M$ for all $y \in \overline{B}$. It is a standard fact that there exists a constant $c > 0$ such that for every $x \in \overline{B}$ it holds that $\mathbb{P}^x(\tau_B \leq t) \leq ct$, for all $t > 0$. Therefore, $0 \leq h(x) - P_t^D h(x) \leq Mct$, for all $x \in \overline{B}$ and all $t > 0$. It follows that for every $x \in \overline{B}$,

$$
\left| \int_0^\epsilon (P_t^D h - h)(x) \, du(t) \right| \leq Mc \left| \int_0^\epsilon t \, du(t) \right|.
$$

By use of (5.14) we get that

$$
\lim_{\epsilon \downarrow 0} \int_0^\epsilon (P_t^D h(x) - h(x)) \, du(t) = 0
$$

uniformly on $\overline{B}$. The proof is now complete.                                      □

**Proposition 5.76.** *If $s$ is an excessive function with respect to $X^D$, then*

$$
s(x) = V^D g(x), \quad x \in D,
$$

*where $g$ is the excessive function for $Y^D$ given by the formula*

$$
g(x) = u(\infty)s(x) + \int_0^\infty (P_t^D s(x) - s(x)) \, du(t) \tag{5.113}
$$

$$
= \psi(0)s(x) + \int_0^\infty (s(x) - P_t^D s(x)) \, d\nu(t). \tag{5.114}
$$

**Proof.** We know that the result is true when $s$ is the potential of a measure. Let $s$ be an arbitrary excessive function of $X^D$. By the Riesz decomposition theorem (see, for instance, Chapter 6 of [23]), $s = G^D m + h$, where $m$ is a measure on $D$, and $h$ is a nonnegative harmonic function for $X^D$. By linearity, it suffices to prove the result for nonnegative harmonic functions.

In the rest of the proof we assume therefore that $s$ is a nonnegative harmonic function for $X^D$. Define the function $g$ by formula (5.113). We have to prove that $g$ is excessive for $Y^D$ and $s = V^D g$. By Lemma 5.75, we know that $g$ is continuous.

Further, since $s$ is excessive, there exists a sequence of nonnegative functions $f_n$ such that $s_n := G^D f_n$ increases to $s$. Then also $P_t^D s_n \uparrow P_t^D s$, implying $s_n - P_t^D s_n \to s - P_t^D s$. If

$$g_n = u(\infty)s_n + \int_0^\infty (s_n - P_t^D s_n)(-du(t)),$$

then we know that $s_n = V^D g_n$ and $g_n$ is excessive for $Y^D$. By use of Fatou's lemma we get that

$$g = u(\infty)s + \int_0^\infty (s - P_t^D s)(-du(t))$$
$$= \lim_n u(\infty)s_n + \int_0^\infty \lim_n (s_n - P_t^D s_n)(-du(t))$$
$$\leq \liminf_n \left( u(\infty)s_n + \int_0^\infty (s_n - P_t^D s_n)(-du(t)) \right)$$
$$= \liminf_n g_n .$$

This implies (again by Fatou's lemma) that

$$V^D g \leq V^D (\liminf_n g_n) \tag{5.115}$$
$$\leq \liminf_n V^D g_n = \liminf_n s_n = s .$$

For any nonnegative function $f$, put $G_1^D f(x) := \int_0^\infty e^{-t} P_t^D f(x)\, dt$, and define $s^1 := s - G_1^D s$. Using an argument similar to that of the proof of Proposition 5.73 we can show that $G^D s$ is not identically infinite. Thus by the resolvent equation we get $G^D s^1 = G^D s - G^D G_1^D s = G_1^D s$, or equivalently,

$$s(x) = s^1(x) + G_1^D s(x) = s^1(x) + G^D s^1(x), \quad x \in D,$$

By use of formula (5.110) for the potential $G^D s^1$, Fubini's theorem and the easy fact that $V^D$ and $G_1^D$ commute, we have

$$G_1^D s = G^D s^1 = V^D \left( u(\infty) G^D s^1 + \int_0^\infty (P_t^D G^D s^1 - G^D s^1)\, du(t) \right)$$
$$= V^D \left( u(\infty) G_1^D s + \int_0^\infty (P_t^D G_1^D s - G_1^D s)\, du(t) \right)$$
$$= G_1^D V^D \left( u(\infty)s + \int_0^\infty (P_t^D s - s)\, du(t) \right).$$

By the uniqueness principle it follows that

$$s = V^D \left( u(\infty)s + \int_0^\infty (P_t^D s - s)\, du(t) \right) = V^D g \quad \text{a.e. in } D .$$

Together with (5.115), this implies that $V^D g = V^D(\liminf_n g_n)$ a.e. From Remark 5.74 it follows that

$$g = \liminf_n g_n \quad \text{a.e.} \tag{5.116}$$

By Fatou's lemma and the $Y^D$-excessiveness of $g_n$ we get that,

$$\lambda U_\lambda^D g = \lambda U_\lambda^D(\liminf g_n) \leq \liminf_n \lambda U_\lambda^D g_n \leq \liminf g_n = g \quad \text{a.e.}$$

We want to show that, in fact, $\lambda U_\lambda^D g \leq g$ everywhere, i.e., that $g$ is supermedian. In order to do this we define $\tilde{g} := \sup_{n \in \mathbb{N}} n U_n^D g$. Then $\tilde{g} \leq g$ a.e., hence, by the absolute continuity of $U_n^D$, $n U_n^D \tilde{g} \leq n U_n^D g \leq \tilde{g}$ everywhere. This implies that $\lambda \mapsto \lambda U_\lambda^D \tilde{g}$ is increasing (see, e.g., Lemma 3.6 in [22]), hence $\tilde{g}$ is supermedian. The same argument gives that $n \mapsto n U_n^D g$ is increasing a.e. Define

$$\tilde{\tilde{g}} := \sup_{\lambda > 0} \lambda U_\lambda^D \tilde{g} = \sup_n n U_n^D \tilde{g}.$$

Then $\tilde{\tilde{g}}$ is excessive, and therefore lower semicontinuous. Moreover,

$$\tilde{\tilde{g}} = \sup_n n U_n^D \tilde{g} \leq \tilde{g} \leq g \quad \text{a.e.}$$

Combining this with the continuity of $g$ and the lower semicontinuity of $\tilde{\tilde{g}}$, we can get that $\tilde{\tilde{g}} \leq g$ everywhere. Further, for $x \in D$ such that $\tilde{g}(x) < \infty$, we have by the monotone convergence theorem and the resolvent equation

$$\begin{aligned}
\lambda U_\lambda^D \tilde{g}(x) &= \lim_{n \to \infty} \lambda U_\lambda^D (n U_n^D) g(x) \\
&= \lim_{n \to \infty} \frac{n \lambda}{n - \lambda} (U_\lambda^D g(x) - U_n^D g(x)) \\
&= \lambda U_\lambda^D g(x).
\end{aligned}$$

Since $\tilde{g} < \infty$ a.e., we have

$$\lambda U_\lambda^D \tilde{g} = \lambda U_\lambda^D g \quad \text{a.e.}$$

Together with the definition of $\tilde{g}$ this implies that

$$\tilde{\tilde{g}} = \tilde{g} \quad \text{a.e.} \tag{5.117}$$

By the continuity of $g$ and the fact that the measures $n U_n^D(x, \cdot)$ converge weakly to the point mass at $x$, we have that for every $x \in D$

$$g(x) \leq \liminf_{n \to \infty} n U_n^D g(x) \leq \tilde{g}(x).$$

Hence, by using (5.117), it follows that $g \leq \tilde{\tilde{g}}$ a.e. Since we already proved that $\tilde{\tilde{g}} \leq g$, it holds that $g = \tilde{\tilde{g}}$ a.e. By the absolute continuity of $U_\lambda^D$, $g \geq \tilde{\tilde{g}} \geq \lambda U_\lambda^D \tilde{\tilde{g}} = \lambda U_\lambda^D g$ everywhere, i.e., $g$ is supermedian.

Since it is well known (see e.g. [60]) that a supermedian function which is lower semicontinuous is in fact excessive, this proves that $g$ is excessive for $Y^D$. By Proposition 5.73 we then have that $V^D g \leq s$ is excessive for $X^D$. Moreover, $V^D g = s$ a.e., and both functions being excessive for $X^D$, they are equal everywhere.

It remains to notice that the formula (5.114) follows immediately from (5.113) by noting that $u(\infty) = \psi(0)$ and $du(t) = -d\nu(t)$.                    □

Propositions 5.72 and 5.76 can be combined in the following theorem containing additional information on harmonic functions.

**Theorem 5.77.** *If $s$ is excessive with respect to $X^D$, then there is a function $g$ excessive with respect to $Y^D$ such that $s = V^D g$. The function $g$ is given by the formula (5.110). Furthermore, if $s$ is harmonic with respect to $X^D$, then $g$ is harmonic with respect to $Y^D$.*

*Conversely, if $g$ is excessive with respect to $Y^D$, then the function $s$ defined by $s = V^D g$ is excessive with respect to $X^D$. If, moreover, $g$ is harmonic with respect to $Y^D$, then $s$ is harmonic with respect to $X^D$.*

*Every nonnegative harmonic function for $Y^D$ is continuous.*

**Proof.** It remains to show the statements about harmonic functions. First note that every excessive functions $g$ for $Y^D$ admits the Riesz decomposition $g = U^D m + h$ where $m$ is a Borel measure on $D$ and $h$ is harmonic function of $Y^D$ (see Chapter 6 of [23] and note that the assumptions on pp. 265, 266 are satisfied). We have already mentioned that excessive functions of $X^D$ admit such decomposition. Since excessive functions of $X^D$ and $Y^D$ are in 1-1 correspondence, and since potentials of measures of $X^D$ and $Y^D$ are in 1-1 correspondence, the same must hold for nonnegative harmonic functions of $X^D$ and $Y^D$.

The continuity of nonnegative harmonic functions for $Y^D$ follows from Lemma 5.75 and Proposition 5.76.                    □

It follows from the theorem above that $V^D$ is a bijection from $\mathcal{S}(Y^D)$ to $\mathcal{S}(X^D)$, and is also a bijection from $\mathcal{H}^+(Y^D)$ to $\mathcal{H}^+(X^D)$. We are going to use $(V^D)^{-1}$ to denote the inverse map and so we have for any $s \in \mathcal{S}(Y^D)$,

$$(V^D)^{-1} s(x) = u(\infty) s(x) + \int_0^\infty (P_t^D s(x) - s(x))\, du(t) \qquad (5.118)$$

$$= \psi(0) s(x) + \int_0^\infty (s(x) - P_t^D s(x))\, d\nu(t).$$

Although the map $V^D$ is order preserving, we do not know if the inverse map $(V^D)^{-1}$ is order preserving on $\mathcal{S}(X^D)$. However from the formula above we can see that $(V^D)^{-1}$ is order preserving on $\mathcal{H}^+(X^D)$.

By combining Proposition 5.72 and Theorem 5.77 we get the following relation which we are going to use later.

**Proposition 5.78.** *For any $x, y \in D$, we have*

$$U^D(x, y) = (V^D)^{-1}(G^D(\cdot, y))(x).$$

### 5.5.3 Harnack Inequality for Subordinate Process

In this subsection we are going to prove the Harnack inequality for positive harmonic functions for the process $Y^D$ under the assumption that $D$ is a bounded domain such that $(P_t^D)$ is intrinsic ultracontractive. The proof we offer uses the intrinsic ultracontractivity in an essential way, and differs from the existing proofs of Harnack inequalities in other settings.

We first recall that since $(P_t^D : t \geq 0)$ is intrinsic ultracontractive, by Theorem 4.2.5 of [66] there exists $T > 0$ such that

$$\frac{1}{2}e^{-\lambda_0 t}\varphi_0(x)\varphi_0(y) \leq p^D(t, x, y) \leq \frac{3}{2}e^{-\lambda_0 t}\varphi_0(x)\varphi_0(y), \quad t \geq T, \ x, y \in D.$$
$$(5.119)$$

**Lemma 5.79.** *Suppose that $D$ is a bounded domain such that $(P_t^D)$ is intrinsic ultracontractive. There exists a constant $C > 0$ such that*

$$V^D g \leq Cg, \quad \forall g \in \mathcal{S}(Y^D). \tag{5.120}$$

**Proof.** Let $T$ be the constant from (5.119). For any nonnegative function $f$,

$$U^D f(x) = \left( \int_0^T P_t^D f(x)u(t)\, dt + \int_T^\infty P_t^D f(x)u(t)\, dt \right).$$

We obviously have

$$\int_0^T P_t^D f(x)u(t)\, dt \geq u(T) \int_0^T P_t^D f(x)\, dt.$$

By using (5.119) we see that

$$\int_T^\infty P_t^D f(x)u(t)\, dt \geq \left( \frac{1}{2} \int_T^\infty e^{-\lambda_0 t}u(t)\, dt \right) \int_D \varphi_0(x)\varphi_0(y)f(y)\, dy$$

$$= c_1 \int_D \varphi_0(x)\varphi_0(y)f(y)\, dy$$

and

$$\int_T^\infty P_t^D f(x)\, dt \le \left(\frac{3}{2}\int_T^\infty e^{-\lambda_0 t} dt\right)\int_D \varphi_0(x)\varphi_0(y)f(y)\, dy$$
$$= c_2 \int_D \varphi_0(x)\varphi_0(y)f(y)\, dy.$$

The last two displays imply that

$$\int_T^\infty P_t^D f(x)u(t)\, dt \ge \frac{c_1}{c_2}\int_T^\infty P_t^D f(x)\, dt.$$

Therefore,

$$U^D f(x) \ge u(T)\int_0^T P_t^D f(x)\, dt + \frac{c_1}{c_2}\int_T^\infty P_t^D f(x)\, dt$$
$$\ge C\int_0^\infty P_t^D f(x)\, dt = CG^D f(x).$$

From $G^D f(x) = V^D U^D f(x)$, we obtain $V^D U^D f(x) \le CU^D f(x)$. Since every $g \in \mathcal{S}(Y^D)$ is an increasing limit of potentials $U^D f(x)$, the claim follows. $\quad\square$

**Lemma 5.80.** *Suppose $D$ is a bounded domain such that $(P_t^D)$ is intrinsic ultracontractive. If $g \in \mathcal{S}(Y^D)$, then for any $x \in D$,*

$$g(x) \ge \frac{1}{2C}e^{-\lambda_0 T}\frac{1}{\psi(\lambda_0)}\varphi_0(x)\int_D g(y)\varphi_0(y)\, dy,$$

*where $T$ is the constant in (5.119) and $C$ is the constant in (5.120).*

**Proof.** From the lemma above we know that, for every $x \in D$, $V^D g(x) \le Cg(x)$, where $C$ is the constant in (5.120). Since $V^D g$ is in $\mathcal{S}(X^D)$, we have

$$V^D g(x) \ge \int_D p^D(T, x, y)V^D g(y)\, dy$$
$$\ge \frac{1}{2}e^{-\lambda_0 T}\varphi_0(x)\int_D \varphi_0(y)V^D g(y)\, dy.$$

Hence

$$Cg(x) \ge V^D g(x) \ge \frac{1}{2}e^{-\lambda_0 T}\varphi_0(x)\int_D \varphi_0(y)V^D g(y)\, dy$$
$$= \frac{1}{2}e^{-\lambda_0 T}\varphi_0(x)\int_D g(y)V^D \varphi_0(y)\, dy$$
$$= \frac{1}{2}e^{-\lambda_0 T}\frac{1}{\psi(\lambda_0)}\varphi_0(x)\int_D g(y)\varphi_0(y)\, dy,$$

where the last line follows from

$$V^D \varphi_0(y) = \int_0^\infty P_t^D \varphi_0(y)\, V(dt) = \int_0^\infty e^{-\lambda_0 t} \varphi_0(y)\, V(dt)$$

$$= \varphi_0(y)\mathcal{L}V(\lambda_0) = \frac{\varphi_0(y)}{\psi(\lambda_0)}.$$

□

In particular, it follows from the lemma that if $g \in \mathcal{S}(Y^D)$ is not identically infinite, then $\int_D \varphi_0(y)g(y)\, dy < \infty$.

**Theorem 5.81.** *Suppose $D$ is a bounded domain such that $(P_t^D)$ is intrinsic ultracontractive. For any compact subset $K$ of $D$, there exists a constant $C$ depending on $K$ and $D$ such that for any $h \in \mathcal{H}^+(Y^D)$,*

$$\sup_{x \in K} h(x) \leq C \inf_{x \in K} h(x).$$

**Proof.** If the conclusion of the theorem were not true, for any $n \geq 1$, there would exist $h_n \in \mathcal{H}^+(Y^D)$ such that

$$\sup_{x \in K} h_n(x) \geq n2^n \inf_{x \in K} h_n(x). \qquad (5.121)$$

By the lemma above, we may assume without loss of generality that

$$\int_D h_n(y)\varphi_0(y)dy = 1, \quad n \geq 1.$$

Define

$$h(x) = \sum_{n=1}^\infty 2^{-n} h_n(x), \quad x \in D.$$

Then

$$\int_D h(y)\varphi_0(y)dy = 1,$$

and so $h \in \mathcal{H}^+(Y^D)$. By (5.121) and the lemma above, for every $n \geq 1$, there exists $x_n \in K$ such that $h_n(x_n) \geq n2^n c_1$ where

$$c_1 = \frac{1}{2C} e^{-\lambda_0 T} \frac{1}{\psi(\lambda_0)} \inf_{x \in K} \varphi_0(x)$$

with $T$ as in (5.119) and $C$ in (5.120). Therefore we have $h(x_n) \geq nc_1$. Since $K$ is compact, there is a convergent subsequence of $x_n$. Let $x_0$ be the limit of this convergent subsequence. Theorem 5.77 implies that $h$ is continuous, and so we have $h(x_0) = \infty$. This is a contradiction. So the conclusion of the theorem is valid.

□

### 5.5.4 *Martin Boundary of Subordinate Process*

In this subsection we assume that $d \geq 3$ and that $D$ is a bounded Lipschitz domain in $\mathbb{R}^d$. Fix a point $x_0 \in D$ and set

$$M^D(x,y) = \frac{G^D(x,y)}{G^D(x_0,y)}, \quad x,y \in D.$$

It is well known that the limit $\lim_{D \ni y \to z} M^D(x,y)$ exists for every $x \in D$ and $z \in \partial D$. The function $M^D(x,z) := \lim_{D \ni y \to z} M^D(x,y)$ on $D \times \partial D$ defined above is called the Martin kernel of $X^D$ based at $x_0$. The Martin boundary and minimal Martin boundary of $X^D$ both coincide with the Euclidean boundary $\partial D$. For these and other results about the Martin boundary of $X^D$ one can see [10]. One of the goals of this section is to determine the Martin boundary of $Y^D$.

By using the Harnack inequality, one can easily show that (see, for instance, pages 17–18 of [71]), if $(h_j)$ is a sequence of functions in $\mathcal{H}^+(X^D)$ converging pointwise to a function $h \in \mathcal{H}^+(X^D)$, then $(h_j)$ is locally uniformly bounded in $D$ and equicontinuous at every point in $D$. Using this, one can get that, if $(h_j)$ is a sequence of functions in $\mathcal{H}^+(X^D)$ converging pointwise to a function $h \in \mathcal{H}^+(X^D)$, then $(h_j)$ converges to $h$ uniformly on compact subsets of $D$. We are going to use this fact below.

**Lemma 5.82.** *Suppose that $x_0 \in D$ is a fixed point.*

*(a)  Let $(x_j : j \geq 1)$ be a sequence of points in $D$ converging to $x \in D$ and let $(h_j)$ be a sequence of functions in $\mathcal{H}^+(X^D)$ with $h_j(x_0) = 1$ for all $j$. If the sequence $(h_j)$ converges to a function $h \in \mathcal{H}^+(X^D)$, then for each $t > 0$*

$$\lim_{j \to \infty} P_t^D h_j(x_j) = P_t^D h(x).$$

*(b)  If $(y_j : j \geq 1)$ is a sequence of points in $D$ such that $\lim_j y_j = z \in \partial D$, then for each $t > 0$ and for each $x \in D$*

$$\lim_{j \to \infty} P_t^D \left( \frac{G^D(\cdot, y_j)}{G^D(x_0, y_j)} \right)(x) = P_t^D(M^D(\cdot, z))(x).$$

**Proof.** (a) For each $j \in \mathbb{N}$, since $h_j(x_0) = 1$, there exists a probability measure $\mu_j$ on $\partial D$ such that

$$h_j(x) = \int_{\partial D} M^D(x,z)\mu_j(dz), \quad x \in D.$$

Similarly, there exists a probability measure $\mu$ on $\partial D$ such that

$$h(x) = \int_{\partial D} M^D(x,z)\mu(dz), \quad x \in D.$$

Let $D_0$ be a relatively compact open subset of $D$ such that $x_0 \in D_0$, and also $x, x_j \in D_0$. Then

$$|P_t^D h_j(x_j) - P_t^D h(x)|$$

$$= \left| \int_D p^D(t, x_j, y) h_j(y) dy - \int_D p^D(t, x, y) h(y) dy \right|$$

$$\leq \left| \int_{D_0} p^D(t, x_j, y) h_j(y) dy - \int_{D_0} p^D(t, x, y) h(y) dy \right|$$

$$+ \int_{D \backslash D_0} p^D(t, x_j, y) h_j(y) \, dy + \int_{D \backslash D_0} p^D(t, x, y) h(y) \, dy \,.$$

Recall that (see Section 6.2 of [62], for instance) there exists a constant $c > 0$ such that

$$\frac{G^D(x, y) G^D(y, w)}{G^D(x, w)} \leq c \left( \frac{1}{|x - y|^{d-2}} + \frac{1}{|y - w|^{d-2}} \right), \quad x, y, w \in D. \quad (5.122)$$

From this and the definition of the Martin kernel we immediately get

$$G^D(x_0, y) M^D(y, z) \leq c \left( \frac{1}{|x_0 - y|^{d-2}} + \frac{1}{|y - z|^{d-2}} \right), \quad y \in D, z \in \partial D.$$
$$(5.123)$$

Recall (see [66], p.131, Theorem 4.6.11) that there is a constant $c > 0$ such that

$$\varphi_0(x_0) \varphi_0(y) \leq c \, G^D(x_0, y), \quad y \in D \,.$$

By the boundedness of $\varphi_0$ we have that $\varphi_0(u) \leq c_1 \varphi_0(x_0)$ for every $u \in D$. Hence it follows from the last display that

$$\varphi_0(u) \varphi_0(y) \leq c \, G^D(x_0, y), \quad u, y \in D, \quad (5.124)$$

with a possibly different constant $c > 0$. Now using (5.109), (5.123) and (5.124) we get that for any $u \in D$,

$$\int_{D \backslash D_0} p^D(t, u, y) h(y) \, dy \leq c_t \varphi_0(u) \int_{D \backslash D_0} \varphi_0(y) h(y) \, dy$$

$$= c_t \varphi_0(u) \int_{D \backslash D_0} dy \, \varphi_0(y) \int_{\partial D} M^D(y, z) \mu(dz)$$

$$= c_t \varphi_0(u) \int_{\partial D} \mu(dz) \int_{D \backslash D_0} \varphi_0(y) M^D(y, z) \, dy$$

$$\leq c c_t \int_{\partial D} \mu(dz) \int_{D \backslash D_0} G^D(x_0, y) M^D(y, z) \, dy$$

$$\leq cc_t \int_{\partial D} \mu(dz) \int_{D \setminus D_0} \left( \frac{1}{|y-z|^{d-2}} + \frac{1}{|x_0 - y|^{d-2}} \right) dy$$

$$\leq cc_t \int_{\partial D} \mu(dz) \int_{D \setminus D_0} \sup_{z \in \partial D} \left( \frac{1}{|y-z|^{d-2}} + \frac{1}{|x_0 - y|^{d-2}} \right) dy$$

$$= cc_t \int_{D \setminus D_0} \sup_{z \in \partial D} \left( \frac{1}{|y-z|^{d-2}} + \frac{1}{|x_0 - y|^{d-2}} \right) dy .$$

The same estimate holds with $h_j$ instead of $h$. For a given $\epsilon > 0$ choose $D_0$ large enough so that the last line in the display above is less than $\epsilon$. Put $A = \sup_{D_0} h$. Take $j_0 \in \mathbb{N}$ large enough so that for all $j \geq j_0$ we have

$$|p^D(t, x_j, y) - p^D(t, x, y)| \leq \epsilon \quad \text{and} \quad |h_j(y) - h(y)| < \epsilon$$

for all $y \in D_0$. Then

$$\left| \int_{D_0} p^D(t, x_j, y) h_j(y) dy - \int_{D_0} p^D(t, x, y) h(y) dy \right|$$

$$\leq \int_{D_0} p^D(t, x_j, y) |h_j(y) - h(y)| dy + \int_{D_0} |p^D(t, x_j, y) - p^D(t, x, y)| h(y) dy$$

$$\leq \epsilon + A|D_0|\epsilon,$$

where $|D_0|$ stands for the Lebesgue measure of $D_0$. This proves the first part.
(b) We proceed similarly as in the proof of the first part. The only difference is that we use (5.122) to get the following estimate:

$$\int_{D \setminus D_0} p^D(t, x, y) \frac{G^D(y, y_j)}{G^D(x_0, y_j)} dy$$

$$\leq c_t \varphi_0(x) \int_{D \setminus D_0} \varphi_0(y) \frac{G^D(y, y_j)}{G^D(x_0, y_j)} dy$$

$$\leq cc_t \int_{D \setminus D_0} \frac{G^D(x_0, y) G^D(y, y_j)}{G^D(x_0, y_j)} dy$$

$$\leq cc_t \int_{D \setminus D_0} \left( |x_0 - y|^{2-d} + |y - y_j|^{2-d} \right) dy$$

$$\leq cc_t \sup_j \int_{D \setminus D_0} \left( |x_0 - y|^{2-d} + |y - y_j|^{2-d} \right) dy .$$

The corresponding estimate for $M^D(\cdot, z)$ is given in part (a) of the lemma. For a given $\epsilon > 0$ find $D_0$ large enough so that the last line in the display above is less than $\epsilon$. Then find $j_0 \in \mathbb{N}$ such that for all $j \geq j_0$,

$$\left| \frac{G^D(y, y_j)}{G^D(x_0, y_j)} - M^D(y, z) \right| < \epsilon, \quad y \in D_0.$$

Then

$$\int_{D_0} p^D(t,x,y) \left| \frac{G^D(y,y_n)}{G^D(x_0,y_j)} - M^D(y,z) \right| dy < \epsilon \quad \text{for all } j \geq j_0.$$

This proves the second part.                                                    □

**Theorem 5.83.** *Suppose that $D \subset \mathbb{R}^d$, $d \geq 3$, is a bounded Lipschitz domain and let $x_0 \in D$ be a fixed point.*

(a)  *If $(x_j)$ is a sequence of points in $D$ converging to $x \in D$ and $(h_j)$ is a sequence of functions in $\mathcal{H}^+(X^D)$ converging to a function $h \in \mathcal{H}^+(X^D)$, then*
$$\lim_j (V^D)^{-1} h_j(x_j) = (V^D)^{-1} h(x).$$

(b)  *If $(y_j)$ is a sequence of points in $D$ converging to $z \in \partial D$, then for every $x \in D$,*
$$\lim_j (V^D)^{-1} \Big( \frac{G^D(\cdot, y_j)}{G^D(x_0, y_j)} \Big)(x) = \lim_j \frac{(V^D)^{-1}(G^D(\cdot, y_j))(x)}{G^D(x_0, y_j)}$$
$$= (V^D)^{-1} M^D(\cdot, z)(x).$$

**Proof.** (a) Normalizing by $h_j(x_0)$ if necessary, we may assume without loss of generality that $h_j(x_0) = 1$ for all $j \geq 1$. Let $\epsilon > 0$. By (5.118) we have

$$|(V^D)^{-1} h_j(x_j) - (V^D)^{-1} h(x)|$$
$$= \left| \int_0^\infty (P_t^D h_j(x_j) - h_j(x_j)) \, du(t) - \int_0^\infty (P_t^D h(x) - h(x)) \, du(t) \right.$$
$$\left. + u(\infty)(h_j(x_j) - h(x)) \right|$$
$$\leq \int_0^\epsilon (P_t^D h_j(x_j) - h_j(x_j)) \, du(t) + \int_0^\epsilon (P_t^D h(x) - h(x)) \, du(t)$$
$$+ \left| \int_\epsilon^\infty (P_t^D h_j(x_j) - h_j(x_j)) \, du(t) - \int_\epsilon^\infty (P_t^D h(x) - h(x)) \, du(t) \right|$$
$$+ u(\infty)|h_j(x_j) - h(x)|.$$

The last term clearly converges to zero as $j \to \infty$.

For any $x \in D$ choose $r > 0$ such that $B(x, 2r) \subset D$ and put $B = B(x,r)$. Without loss of generality we may and do assume that $x_j \in B$ for all $j \geq 1$. Since $h$ and $h_j$ are continuous in $D$ and $(h_j)$ is locally uniformly bounded in $D$, there is a constant $M > 0$ such that $h$ and $h_j$, $j = 1, 2, \ldots$, are all bounded from above by $M$ on $\overline{B}$. Now from the proof of Lemma 5.75, more precisely from display (5.112), it follows that there is a constant $c_1 > 0$ such that

$$0 \leq h(y) - P_t^D h(y) \leq c_1 t, \quad y \in \overline{B},$$

and

$$0 \le h_j(y) - P_t^D h_j(y) \le c_1 t, \quad y \in \overline{B}, j \ge 1.$$

Therefore we have,

$$\left| \int_0^\epsilon (P_t^D h - h)(y)\, du(t) \right| \le c_1 \left| \int_0^\epsilon t\, du(t) \right|, \quad y \in \overline{B}$$

and

$$\left| \int_0^\epsilon (P_t^D h_j - h_j)(y)\, du(t) \right| \le c_1 \left| \int_0^\epsilon t\, du(t) \right|, \quad y \in \overline{B}, j \ge 1.$$

Using (5.14) we get that

$$\lim_{\epsilon \downarrow 0} \int_0^\epsilon (P_t^D h(x) - h(x))\, du(t) = 0,$$

and

$$\lim_{\epsilon \downarrow 0} \int_0^\epsilon (P_t^D h_j(x_j) - h_j(x_j))\, du(t) = 0.$$

Further,

$$\left| \int_\epsilon^\infty (P_t^D h_j(x_j) - h_j(x_j))\, du(t) - \int_\epsilon^\infty (P_t^D h(x) - h(x))\, du(t) \right|$$
$$\le \int_\epsilon^\infty (|h_j(x_j) - h(x_j)| + |h(x_j) - h(x)|)\, du(t) + \int_\epsilon^\infty |P_t^D h_j(x_j) - P_t^D h(x)|\, du(t).$$

Since $|h_j(x_j) - h(x_j)| + |h(x_j) - h(x)| \le 2M$ and $|P_t^D h_j(x_j) - P_t^D h(x)| \le M$ for all $j \ge 1$ and all $x \in \overline{B}$, we can apply Lemma 5.82(a) and the dominated convergence theorem to get

$$\lim_{j \to \infty} \int_\epsilon^\infty (|h_j(x_j) - h(x_j)| + |h(x_j) - h(x)|)\, du(t) = 0$$

and

$$\lim_{j \to \infty} \int_\epsilon^\infty |P_t^D h_j(x_j) - P_t^D h(x)|\, du(t) = 0.$$

The proof of (a) is now complete.

(b) The proof of (b) is similar to (a). The only difference is that we use 5.82(b) in this case. We omit the details. □

Let us define the function $K_Y^D(x, z) := (V^D)^{-1} M^D(\cdot, z)(x)$ on $D \times \partial D$. For each fixed $z \in \partial D$, $K_Y^D(\cdot, z) \in \mathcal{H}^+(Y^D)$. By the first part of Theorem 5.83, we know that $K_Y^D(x, z)$ is continuous on $D \times \partial D$. Let $(y_j)$ be a sequence of points in $D$ converging to $z \in \partial D$, then from Theorem 5.83(b) we get that

$$K_Y^D(x, z) = \lim_{j \to \infty} (V^D)^{-1} \left( \frac{G^D(\cdot, y_j)}{G^D(x_0, y_j)} \right)(x)$$

$$= \lim_{j \to \infty} \frac{(V^D)^{-1}(G^D(\cdot, y_j))(x)}{G^D(x_0, y_j)}$$

$$= \lim_{j \to \infty} \frac{U^D(x, y_j)}{G^D(x_0, y_j)}, \tag{5.125}$$

where the last line follows from Proposition 5.78. In particular, there exists the limit

$$\lim_{j \to \infty} \frac{U^D(x_0, y_j)}{G^D(x_0, y_j)} = K_Z^D(x_0, z). \tag{5.126}$$

Now we define a function $M_Y^D$ on $D \times \partial D$ by

$$M_Y^D(x, z) := \frac{K_Y^D(x, z)}{K_Y^D(x_0, z)}, \quad x \in D, z \in \partial D. \tag{5.127}$$

For each $z \in \partial D$, $M_Y^D(\cdot, z) \in \mathcal{H}_+(Y^D)$. Moreover, $M_Y^D$ is jointly continuous on $D \times \partial D$. From the definition above and (5.125) we can easily see that

$$\lim_{D \ni y \to z} \frac{U^D(x, y)}{U^D(x_0, y)} = M_Y^D(x, z), \quad x \in D, z \in \partial D. \tag{5.128}$$

**Theorem 5.84.** *Let $D \subset \mathbb{R}^d$, $d \geq 3$, be a bounded Lipschitz domain. The Martin boundary and the minimal Martin boundary of $Y^D$ both coincide with the Euclidean boundary $\partial D$, and the Martin kernel based at $x_0$ is given by the function $M_Y^D$.*

**Proof.** The fact that $M_Y^D$ is the Martin kernel of $Y^D$ based at $x_0$ has been proven in the paragraph above. It follows from Theorem 5.77 that when $z_1$ and $z_2$ are two distinct points on $\partial D$, the functions $M_Y^D(\cdot, z_1)$ and $M_Y^D(\cdot, z_2)$ are not identical. Therefore the Martin boundary of $Y^D$ coincides with the Euclidean boundary $\partial D$. Since $M^D(\cdot, z) \in \mathcal{H}^+(X^D)$ is minimal, by the order preserving property of $(V^D)^{-1}$ we know that $M_Y^D(\cdot, z) \in \mathcal{H}^+(Y^D)$ is also minimal. Therefore the minimal Martin boundary of $Y_D$ also coincides with the Euclidean boundary $\partial D$.                                                    □

It follows from Theorem 5.84 and the general theory of Martin boundary that for any $g \in \mathcal{H}^+(Y^D)$ there exists a finite measure $n$ on $\partial D$ such that

$$g(x) = \int_{\partial D} M_Y^D(x, z) n(dz), \quad x \in D.$$

The measure $n$ is sometimes called the Martin measure of $g$. The following result gives the relation between the Martin measure of $h \in \mathcal{H}^+(X^D)$ and the Martin measure of $(V^D)^{-1}h \in \mathcal{H}^+(Y^D)$.

**Proposition 5.85.** *If $h \in \mathcal{H}^+(X^D)$ has the representation*

$$h(x) = \int_{\partial D} M^D(x, z) \, m(dz), \quad x \in D,$$

*then*

$$(V^D)^{-1}h(x) = \int_{\partial D} M_Y^D(x, z) \, n(dz), \quad x \in D$$

*with* $n(dz) = K_Y^D(x_0, z) \, m(dz)$.

**Proof.** By assumption we have

$$h(x) = \int_{\partial D} M^D(x, z) \, m(dz), \quad x \in D.$$

Using (5.113) and Fubini's theorem we get

$$(V^D)^{-1}h(x) = \int_{\partial D} (V^D)^{-1}(M^D(\cdot, z))(x) \, m(dz)$$

$$= \int_{\partial D} M_Y^D(x, z) K_Y^D(x_0, z) \, m(dz) = \int_{\partial D} M_Y^D(x, z) \, n(dz),$$

with $n(dz) = K_Y^D(x_0, z) m(dz)$. The proof is now complete.  $\square$

From Theorem 5.83 we know that $(V^D)^{-1} : \mathcal{H}^+(X^D) \to \mathcal{H}^+(Y^D)$ is continuous with respect to topologies of locally uniform convergence. In the next result we show that $V^D : \mathcal{H}^+(Y^D) \to \mathcal{H}^+(X^D)$ is also continuous.

**Proposition 5.86.** *Let $(g_j, j \geq 0)$ be a sequence of functions in $\mathcal{H}^+(Y^D)$ converging pointwise to the function $g \in \mathcal{H}^+(Y^D)$. Then $\lim_{j \to \infty} V^D g_j(x) = V^D g(x)$ for every $x \in D$.*

**Proof.** Without loss of generality we may assume that $g_j(x_0) = 1$ for all $j \in \mathbb{N}$. Then there exist probability measures $n_j$, $j \in \mathbb{N}$, and $n$ on $\partial D$ such that $g_j(x) = \int_{\partial D} M_Y^D(x, z) n_j(dz)$, $j \in \mathbb{N}$, and $g(x) = \int_{\partial D} M_Y^D(x, z) n(dz)$. It is easy to show that the convergence of the harmonic functions $h_j$ implies that $n_j \to n$ weakly. Let $V^D g_j(x) = \int_{\partial D} M^D(x, z) m_j(dz)$ and $V^D g(x) = \int_{\partial D} M^D(x, z) m(dz)$. Then $n_j(dz) = K_Y^D(x_0, z) m_j(dz)$ and $n(dz) = K_Y^D(x_0, z) m(dz)$. Since the density $K_Y^D(x_0, \cdot)$ is bounded away from zero and bounded from above, it follows that $m_j \to m$ weakly. From this the claim of proposition follows immediately.  $\square$

### 5.5.5  Boundary Harnack Principle for Subordinate Process

The boundary Harnack principle is a very important result in potential theory and harmonic analysis. For example, it is usually used to prove that, when $D$ is a bounded Lipschitz domain, both the Martin boundary and the minimal Martin boundary of $X^D$ coincide with the Euclidean boundary $\partial D$. We have already proved in Theorem 5.84 that for $Y^D$, both the Martin boundary and the minimal Martin boundary coincide with the Euclidean boundary $\partial D$. By using this we are going to prove a boundary Harnack principle for functions in $\mathcal{H}^+(Y^D)$.

In this subsection we will always assume that $D \subset \mathbb{R}^d$, $d \geq 3$, is a bounded Lipschitz domain and $x_0 \in D$ is fixed. Recall that $\varphi_0$ is the eigenfunction corresponding to the smallest eigenvalue $\lambda_0$ of $-\Delta|_D$. Also recall that the potential operator $V^D$ is not absolutely continuous in case $b > 0$ and is given by

$$V^D f(x) = bf(x) + \int_0^\infty P_t^D f(x)v(t)\,dt\,.$$

Define

$$\tilde{V}^D(x,y) = \int_0^\infty p^D(t,x,y)v(t)\,dt\,.$$

Then

$$V^D f(x) = bf(x) + \int_D \tilde{V}^D(x,y)f(y)\,dy\,.$$

**Proposition 5.87.** *Suppose that $D$ is a bounded Lipschitz domain. There exist $c > 0$ and $k > d$ such that*

$$U^D(x,y) \leq c\frac{\varphi_0(x)\varphi_0(y)}{|x-y|^k}\,,$$

$$\tilde{V}^D(x,y) \leq c\frac{\varphi_0(x)\varphi_0(y)}{|x-y|^k}\,,$$

*for all $x,y \in D$.*

**Proof.** We give a proof of the second estimate, the proof of the first being exactly the same. Note that similarly as in (2.13)

$$\lim_{t\to 0} tv(t) = 0\,. \tag{5.129}$$

It follows from Theorem 4.6.9 of [66] that the density $p^D$ of the killed Brownian motion on $D$ satisfies the following estimate

$$p^D(t,x,y) \leq c_1 t^{-k/2}\varphi_0(x)\varphi_0(y)e^{-\frac{|x-y|^2}{6t}}\,, \quad t > 0,\ x,y \in D,$$

for some $k > d$ and $c_2 > 0$. Recall that $v$ is a decreasing function. From (5.129) it follows that there exists a $t_0 > 0$ such that $v(t) \leq \frac{1}{t}$ for $t \leq t_0$. Consequently,

$$v(t) \leq M + \frac{1}{t}\,, \quad t > 0,$$

for some $M > 0$. Now we have

$$\tilde{V}^D(x,y) = \int_0^\infty p^D(t,x,y)v(t)dt \le c_1 \int_0^\infty t^{-k/2}\varphi_0(x)\varphi_0(y)e^{-\frac{|x-y|^2}{6t}}v(t)dt$$

$$\le c_1 \int_0^\infty t^{-k/2-1}\varphi_0(x)\varphi_0(y)e^{-\frac{|x-y|^2}{6t}}dt + Mc_1 \int_0^\infty t^{-k/2}\varphi_0(x)\varphi_0(y)e^{-\frac{|x-y|^2}{6t}}dt$$

$$\le c_2 \frac{\varphi_0(x)\varphi_0(y)}{|x-y|^k} + Mc_3 \frac{\varphi_0(x)\varphi_0(y)}{|x-y|^{k-2}}$$

$$\le c_4 \frac{\varphi_0(x)\varphi_0(y)}{|x-y|^k}.$$

The proof is now finished. $\qquad\qquad\qquad\qquad\qquad\qquad\qquad\qquad\qquad\square$

**Lemma 5.88.** *Suppose that $D$ is a bounded Lipschitz domain and $W$ an open subset of $\mathbb{R}^d$ such that $W \cap \partial D$ is non-empty. If $h \in \mathcal{H}^+(Y^D)$ satisfies*

$$\lim_{x \to z} \frac{h(x)}{(V^D)^{-1}1(x)} = 0, \quad \text{for all } z \in W \cap \partial D,$$

*then*

$$\lim_{x \to z} V^D h(x) = 0, \quad \text{for all } z \in W \cap \partial D.$$

**Proof.** Fix $z \in W \cap \partial D$. For any $\epsilon > 0$, there exists $\delta > 0$ such that $h(x) \le \epsilon(V^D)^{-1}1(x)$ for $x \in B(z,\delta) \cap D$. Thus we have

$$V^D h(x) \le V^D(h\mathbf{1}_{D\setminus B(z,\delta)})(x) + \epsilon V^D(V^D)^{-1}1(x)$$
$$= V^D(h\mathbf{1}_{D\setminus B(z,\delta)})(x) + \epsilon, \quad x \in D.$$

For any $x \in B(z,\delta/2) \cap D$ we have

$$V^D(h\mathbf{1}_{D\setminus B(z,\delta)})(x) = bh(x)\mathbf{1}_{D\setminus B(z,\delta)}(x) + \int_{D\setminus B(z,\delta)} \tilde{V}^D(x,y)h(y)\,dy$$

$$= \int_{D\setminus B(z,\delta)} \tilde{V}^D(x,y)h(y)\,dy$$

since $\mathbf{1}_{D\setminus B(z,\delta)}(x) = 0$ for $x \in B(z,\delta/2) \cap D$. By Proposition 5.87 we get that there exists $c > 0$ such that for any $x \in B(z,\delta/2) \cap D$,

$$\int_{D\setminus B(z,\delta)} \tilde{V}^D(x,y)h(y) \le c\varphi_0(x) \int_{D\setminus B(z,\delta)} \frac{\varphi_0(y)}{|x-y|^k}h(y)\,dy$$

$$\le c\varphi_0(x) \int_{D\setminus B(z,\delta)} \frac{\varphi_0(y)}{(\delta/2)^k}h(y)\,dy \le c\varphi_0(x) \int_D \varphi_0(y)h(y)\,dy.$$

Hence,

$$V^D h(x) \leq c\varphi_0(x) \int_D \varphi_0(y) h(y) dy + \epsilon.$$

From Lemma 5.80 we know that $\int_D \varphi_0(y) h(y) dy < \infty$. Now the conclusion of the lemma follows easily from the fact that $\lim_{x \to z} \varphi_0(x) = 0$. $\quad\square$

Now we can prove the main result of this section: the boundary Harnack principle.

**Theorem 5.89.** *Suppose that $D \subset \mathbb{R}^d$, $d \geq 3$, is a bounded Lipschitz domain, $W$ an open subset of $\mathbb{R}^d$ such that $W \cap \partial D$ is non-empty, and $K$ a compact subset of $W$. There exists a constant $c > 0$ such that for any two functions $h_1$ and $h_2$ in $\mathcal{H}^+(Y^D)$ satisfying*

$$\lim_{x \to z} \frac{h_i(x)}{(V^D)^{-1} 1(x)} = 0, \quad z \in W \cap \partial D, \ i = 1, 2,$$

*we have*

$$\frac{h_1(x)}{h_2(x)} \leq c \frac{h_1(y)}{h_2(y)}, \quad x, y \in K \cap D.$$

**Proof.** By use of (5.119) and Proposition 5.87 there exist positive constants $c_1$ and $c_2$ such that

$$c_1 \varphi_0(x) \varphi_0(y) \leq U^D(x,y) \leq c_2 \frac{\varphi_0(x)\varphi_0(y)}{|x-y|^k}, \quad x, y \in D,$$

where $k > d$ is given in Proposition 5.87. Therefore it follows from (5.128) that there exist positive constants $c_3$ and $c_4$ such that

$$c_3 \varphi_0(x) \leq M_Y^D(x,z) \leq c_4 \varphi_0(x), \quad x \in K \cap D, z \in \partial D \setminus W. \qquad (5.130)$$

Suppose that $h_1$ and $h_2$ are two functions in $\mathcal{H}^+(Y^D)$ such that

$$\lim_{x \to z} \frac{h_i(x)}{(V^D)^{-1} 1(x)} = 0, \quad z \in W \cap \partial D, i = 1, 2,$$

then by Lemma 5.88 we have

$$\lim_{x \to z} V^D h_i(x) = 0, \quad z \in W \cap \partial D, i = 1, 2.$$

Now by Corollary 8.1.6 of [128] we know that the Martin measures $m_1$ and $m_2$ of $V^D h_1$ and $V^D h_2$ are supported by $\partial D \setminus W$ and so we have

$$V^D h_i(x) = \int_{\partial D \setminus W} M^D(x,z) \, m_i(dz), \quad x \in D, i = 1, 2.$$

Using Proposition 5.85 we get that

$$h_i(x) = \int_{\partial D \backslash W} M_Y^D(x, z)\, n_i(dz), \quad x \in D, \ i = 1, 2,$$

where $n_i(dz) = K_Y^D(x_0, z)\, m_i(dz), i = 1, 2$. Now using (5.130) it follows that

$$c_3\varphi_0(x)n_i(\partial D \backslash W) \le h_i(x) \le c_4\varphi_0(x)n_i(\partial D \backslash W), \quad x \in K \cap D, \ i = 1, 2.$$

The conclusion of the theorem follows immediately.                                    □

From the proof of Theorem 5.89 we can see that the following result is true.

**Proposition 5.90.** *Suppose that $D \subset \mathbb{R}^d$, $d \ge 3$, is a bounded Lipschitz domain and $W$ an open subset of $\mathbb{R}^d$ such that $W \cap \partial D$ is non-empty. If $h \in \mathcal{H}^+(Y^D)$ satisfies*

$$\lim_{x \to z} \frac{h(x)}{(V^D)^{-1}1(x)} = 0, \quad z \in W \cap \partial D,$$

*then*

$$\lim_{x \to z} h(x) = 0, \quad z \in W \cap \partial D.$$

**Proof.** From the proof of Theorem 5.89 we see that the Martin measure $n$ of $h$ is supported by $\partial D \backslash W$ and so we have

$$h(x) = \int_{\partial D \backslash W} M_Y^D(x, z)n(dz), \quad x \in D.$$

For any $z_0 \in W \cap \partial D$, take $\delta > 0$ small enough so that $B(z_0, \delta) \subset \overline{B(z_0, \delta)} \subset W$. Then it follows from (5.130) that

$$c_5\varphi_0(x) \le M_Y^D(x, z) \le c_6\varphi_0(x), \quad x \in B(z_0, \delta) \cap D, z \in \partial D \backslash W,$$

for some positive constants $c_5$ and $c_6$. Thus

$$h(x) \le c_6\varphi_0(x)n(\partial D \backslash W), \quad x \in B(z_0, \delta) \cap D,$$

from which the assertion of the proposition follows immediately.                      □


### 5.5.6   Sharp Bounds for the Green Function and the Jumping Function of Subordinate Process

In this subsection we are going to derive sharp bounds for the Green function and the jumping function of the process $Y^D$. The method uses the upper and lower bounds for the transition densities $p^D(t, x, y)$ of the killed Brownian motion. The lower bound that we need is available only in case when $D$ is a bounded $C^{1,1}$ domain in $\mathbb{R}^d$. Therefore, throughout this subsection we assume that $D \subset \mathbb{R}^d$ is a bounded $C^{1,1}$ domain. Moreover, recall the standing assumption that $S$ is a special subordinator such that $b > 0$ or $\mu(0,\infty) = \infty$ which guarantees the existence of a decreasing potential density $u$.

Recall that a bounded domain $D \subset \mathbb{R}^d$, $d \geq 2$, is called a bounded $C^{1,1}$ domain if there exist positive constants $r_0$ and $M$ with the following property: For every $z \in \partial D$ and every $r \in (0, r_0]$, there exist a function $\Gamma_z : \mathbb{R}^{d-1} \to \mathbb{R}$ satisfying the condition $|\nabla \Gamma_z(\xi) - \nabla \Gamma_z(\eta)| \leq M|\xi - \eta|$ for all $\xi, \eta \in \mathbb{R}^{d-1}$, and an orthonormal coordinate system $CS_z$ such that if $y = (y_1, \ldots, y_d)$ in $CS_z$ coordinates, then

$$B(z,r) \cap D = B(z,r) \cap \{y : y_d > \Gamma_z(y_1, \ldots, y_{d-1})\}.$$

When we speak of a bounded $C^{1,1}$ domain in $\mathbb{R}$ we mean a finite open interval.

For any $x \in D$, let $\rho(x)$ denote the distance between $x$ and $\partial D$. We will use the following two bounds for transition densities $p^D(t, x, y)$: There exists a positive constant $c_1$ such that for all $t > 0$ and any $x, y \in D$,

$$p^D(t, x, y) \leq c_1 t^{-d/2-1} \rho(x)\rho(y) \exp\left(-\frac{|x-y|^2}{6t}\right). \tag{5.131}$$

This result (valid also for Lipschitz domains) can be found in [66] (see also [144]). The lower bound was obtained in [158] and [143] and states that for any $A > 0$, there exist positive constants $c_2$ and $c$ such that for any $t \in (0, A]$ and any $x, y \in D$,

$$p^D(t, x, y) \geq c_2 \left(\frac{\rho(x)\rho(y)}{t} \wedge 1\right) t^{-d/2} \exp\left(-\frac{c|x-y|^2}{t}\right). \tag{5.132}$$

Recall that the Green function of $Y^D$ is given by

$$U^D(x, y) = \int_0^\infty p^D(t, x, y)u(t)\, dt,$$

where $u$ is the potential density of the subordinator $S$. Instead of assuming conditions on the asymptotic behavior of the Laplace exponent $\phi(\lambda)$ as $\lambda \to \infty$, we will directly assume the asymptotic behavior of $u(t)$ as $t \to 0+$.
**Assumption A:** (i) There exist constants $c_0 > 0$ and $\beta \in [0, 1]$ with $\beta > 1 - d/2$, and a continuous function $\ell : (0, \infty) \to (0, \infty)$ which is slowly varying at $\infty$ such that

$$u(t) \sim \frac{c_0}{t^\beta \ell(1/t)}, \quad t \to 0+. \tag{5.133}$$

(ii) In the case when $d = 1$ or $d = 2$, there exist constants $c > 0, T > 0$ and $\gamma < d/2$ such that

$$u(t) \leq ct^{\gamma-1}, \quad t \geq T. \tag{5.134}$$

Note that under certain assumptions on the asymptotic behavior of $\phi(\lambda)$ as $\lambda \to \infty$, one can obtain (5.133) and (5.134) for the density $u$.

**Theorem 5.91.** *Suppose that $D$ is a bounded $C^{1,1}$ domain in $\mathbb{R}^d$ and that the potential density $u$ of the special subordinator $S = (S_t : t \geq 0)$ satisfies the Assumption A. Suppose also that there is a function $g : (0, \infty) \to (0, \infty)$ such that*

$$\int_0^\infty t^{d/2-2+\beta} e^{-t} g(t) dt < \infty$$

*and $\xi > 0$ such that $f_{\ell,\xi}(y,t) \leq g(t)$ for all $y, t > 0$, where $f_{\ell,\xi}$ is the function defined before Lemma 5.32 using the $\ell$ in (5.133). Then there exist positive constants $C_1 \leq C_2$ such that for all $x, y \in D$,*

$$C_1 \left( \frac{\rho(x)\rho(y)}{|x-y|^2} \wedge 1 \right) \frac{1}{|x-y|^{d+2\beta-2} \ell(\frac{1}{|x-y|^2})} \leq U^D(x,y)$$

$$\leq C_2 \left( \frac{\rho(x)\rho(y)}{|x-y|^2} \wedge 1 \right) \frac{1}{|x-y|^{d+2\beta-2} \ell(\frac{1}{|x-y|^2})}. \tag{5.135}$$

**Proof.** We start by proving the upper bound. Using the obvious upper bound $p^D(t,x,y) \leq (4\pi t)^{-d/2} \exp(-|x-y|^2/4t)$ and Lemma 5.32 one can easily show that

$$U^D(x,y) \leq c_1 \frac{1}{|x-y|^{d+2\beta-2} \ell\left(\frac{1}{|x-y|^2}\right)}.$$

Now note that (5.131) gives

$$U^D(x,y) \leq c_2 \rho(x)\rho(y) \int_0^\infty t^{-d/2-1} e^{-|x-y|^2/6t} u(t) \, dt.$$

Thus it follows from Lemma 5.32 that

$$U^D(x,y) \leq c_3 \rho(x)\rho(y) \frac{1}{|x-y|^{d+2\beta} \ell\left(\frac{1}{|x-y|^2}\right)}.$$

Now combining the two upper bounds obtained so far we arrive at the upper bound in (5.135).

In order to prove the lower bound, we first recall the following result about slowly varying functions (see [21], p. 22, Theorem 1.5.12):

$$\lim_{\lambda\to\infty}\frac{\ell(t\lambda)}{\ell(\lambda)}=1$$

uniformly in $t\in[a,b]$ where $[a,b]\subset(0,\infty)$. Together with joint continuity of $(t,\lambda)\mapsto\ell(t\lambda)/\ell(\lambda)$, this shows that for a given $\lambda_0>0$ and an interval $[a,b]\subset(0,\infty)$, there exists a positive constant $c(a,b,\lambda_0)$ such that

$$\frac{\ell(t\lambda)}{\ell(\lambda)}\le c(a,b,\lambda_0)\,,\quad a\le t\le b,\lambda\ge\lambda_0\,. \tag{5.136}$$

Now, by (5.132),

$$U^D(x,y)\ge c_4\int_0^A\left(\frac{\rho(x)\rho(y)}{t}\wedge 1\right)t^{-d/2}\exp\left(-\frac{c|x-y|^2}{t}\right)u(t)\,dt\,.$$

Assume $x\ne y$. Let $R$ be the diameter of $D$ and assume that $A$ has been chosen so that $A=R^2$. Then for any $x,y\in D$, $\rho(x)\rho(y)<R^2=A$. The lower bound is proved by considering two separate cases:

(i) $|x-y|^2<2\rho(x)\rho(y)$. In this case we have

$$\begin{aligned}U^D(x,y)&\ge c_4\int_0^{\rho(x)\rho(y)}\left(\frac{\rho(x)\rho(y)}{t}\wedge 1\right)t^{-d/2}\exp\{-c|x-y|^2/t\}u(t)\,dt\\ &\ge c_5|x-y|^{-d+2}\int_{\frac{c|x-y|^2}{\rho(x)\rho(y)}}^{\infty}s^{d/2-2}e^{-s}u(c|x-y|^2/s)\,ds\\ &\ge c_5|x-y|^{-d+2}\int_{2c}^{4c}s^{d/2-2}e^{-s}u(c|x-y|^2/s)\,ds\,. \end{aligned}\tag{5.137}$$

For $2c\le s\le 4c$, we have that $1/4\le c|x-y|^2/s\le 1/2$. Hence, by (5.133), there exists $c_6>0$ such that

$$u\left(\frac{c|x-y|^2}{s}\right)\ge\frac{c_6}{\left(\frac{c|x-y|^2}{s}\right)^\beta\ell\left(\frac{s}{c|x-y|^2}\right)}\,.$$

Further, since $1/|x-y|^2\ge 1/R^2$ for all $x,y\in D$, we can use (5.136) to conclude that there exists $c_7>0$ such that

$$\frac{\ell\left(\frac{1}{|x-y|^2}\right)}{\ell\left(\frac{s}{c|x-y|^2}\right)}\ge c_7\,,\quad 2c\le s\le 4c,\,x,y\in D\,.$$

It follows from (5.137), that

$$U^D(x,y) \geq c_5 |x-y|^{-d+2} \int_{2c}^{4c} s^{d/2-2} e^{-s} \frac{c_6 c_7}{\left(\frac{c|x-y|^2}{s}\right)^\beta \ell\left(\frac{1}{|x-y|^2}\right)} \, ds$$

$$= \frac{c_4}{|x-y|^{d+2\beta-2}\ell(\frac{1}{|x-y|^2})} \int_{2c}^{4c} s^{d/2+\beta-2} e^{-s} \, ds$$

$$= \frac{c_9}{|x-y|^{d+2\beta-2}\ell(\frac{1}{|x-y|^2})} .$$

(ii) $|x-y|^2 \geq 2\rho(x)\rho(y)$. In this case we have

$$U^D(x,y) \geq c_4 \rho(x)\rho(y) \int_{\rho(x)\rho(y)}^{A} t^{-d/2-1} \exp\{-c|x-y|^2/t\} u(t) \, dt$$

$$= c_{10}\rho(x)\rho(y)|x-y|^{-d} \int_{\frac{c|x-y|^2}{A}}^{\frac{c|x-y|^2}{\rho(x)\rho(y)}} s^{d/2-1} e^{-s} u(c|x-y|^2/s) \, ds$$

$$\geq c_{10}\rho(x)\rho(y)|x-y|^{-d} \int_{c}^{2c} s^{d/2-1} e^{-s} u(c|x-y|^2/s) \, ds .$$

The integral above is estimated in the same way as in case (i). It follows that there exists a positive constant $c_{11}$ such that

$$U^D(x,y) \geq c_{10}\rho(x)\rho(y)|x-y|^{-d} \frac{c_{11}}{|x-y|^{2\beta}\ell(\frac{1}{|x-y|^2})}$$

$$= c_{12} \frac{\rho(x)\rho(y)}{|x-y|^{d+2\beta}\ell(\frac{1}{|x-y|^2})} .$$

Combining the two cases above we arrive at the lower bound (5.135).    □

Suppose that the subordinator $S$ has a strictly positive drift $b$ and $d \geq 3$. Then we can take $\beta = 0$ and $\ell = 1$ in the Assumption A, and Theorem 5.91 implies that the Green function $U^D$ of $Y^D$ is comparable to the Green function of $X^D$. Further, if $\phi(\lambda) \sim c_0\lambda^{\alpha/2}$, as $\lambda \to \infty$, $0 < \alpha < 2$, then by (5.28) it follows that the Assumption A holds true with $\beta = 1-\alpha/2$ and $\ell = 1$. In this way we recover a result from [146] saying that under the stated assumption,

$$C_1 \left(\frac{\rho(x)\rho(y)}{|x-y|^2} \wedge 1\right) \frac{1}{|x-y|^{d-\alpha}} \leq U^D(x,y) \leq C_2 \left(\frac{\rho(x)\rho(y)}{|x-y|^2} \wedge 1\right) \frac{1}{|x-y|^{d-\alpha}} .$$

The jumping function $J^D(x,y)$ of the subordinate process $Y^D$ is given by the following formula:

$$J^D(x,y) = \int_0^\infty p^D(t,x,y)\,\mu(dt) .$$

Suppose that $\mu(dt)$ has a decreasing density $\mu(t)$ which satisfies

**Assumption B:** There exist constants $c_0 > 0$, $\beta \in [1, 2]$ and a continuous function $\ell : (0, \infty) \to (0, \infty)$ which is slowly varying at $\infty$ such that such that

$$\mu(t) \sim \frac{c_0}{t^\beta \ell(1/t)}, \quad t \to 0 + . \tag{5.138}$$

Then we have the following result on sharp bounds of $J^D(x, y)$. The proof is similar to the proof of Theorem 5.91, and therefore omitted.

**Theorem 5.92.** *Suppose that $D$ is a bounded $C^{1,1}$ domain in $\mathbb{R}^d$ and that the Lévy density $\mu(t)$ of the subordinator $S = (S_t : t \geq 0)$ exists, is decreasing and satisfies the Assumption B. Suppose also that there is a function $g : (0, \infty) \to (0, \infty)$ such that*

$$\int_0^\infty t^{d/2 - 2 + \beta} e^{-t} g(t) dt < \infty$$

*and $\xi > 0$ such that $f_{\ell, \xi}(y, t) \leq g(t)$ for all $y, t > 0$, where $f_{\ell, \xi}$ is the function defined before Lemma 5.32 using the $\ell$ in (5.138). Then there exist positive constants $C_3 \leq C_4$ such that for all $x, y \in D$*

$$C_3 \left( \frac{\rho(x)\rho(y)}{|x - y|^2} \wedge 1 \right) \frac{1}{|x - y|^{d + 2\beta - 2} \ell(\frac{1}{|x - y|^2})} \leq J^D(x, y)$$

$$\leq C_4 \left( \frac{\rho(x)\rho(y)}{|x - y|^2} \wedge 1 \right) \frac{1}{|x - y|^{d + 2\beta - 2} \ell(\frac{1}{|x - y|^2})} . \tag{5.139}$$

# Bibliography

[1] H. Aikawa, *Boundary Harnack principle and Martin boundary for a uniform domain*, J. Math. Soc. Japan 53 (2001), no. 1, 119–145.

[2] H. Aikawa, *Potential-theoretic characterizations of nonsmooth domains*, Bull. London Math. Soc. 36 (2004), no. 4, 469–482.

[3] A. Ancona, *Principe de Harnack a la frontiere et théorème de Fatou pour un opérateur elliptique dans un domaine lipschitzien*, Ann. Inst. Fourier (Grenoble) 28 (1978), no. 4, 169–213.

[4] D. Armitage, S. Gardiner, *Classical potential theory*, Springer-Verlag, London 2001.

[5] R. Bañuelos, K. Bogdan, *Symmetric stable processes in cones*, Potential Anal. 21 (2004), no. 3, 263–288.

[6] R. Bañuelos, *Intrinsic ultracontractivity and eigenfunction estimates for Schrödinger operators*, J. Funct. Anal. 100 (1991), 181–206.

[7] R. Bañuelos, T. Kulczycki, *The Cauchy process and the Steklov problem*, J. Funct. Anal. 211 (2004), 355–423.

[8] R. Bañuelos, T. Kulczycki, *Spectral gap for the Cauchy process on convex, symmetric domains*. Comm. in Partial Diff. Equations 31 (2006), 1841–1878.

[9] R. Bañuelos, P.J. Méndez-Hernández, *Sharp inequalities for heat kernels of Schrödinger operators and applications to spectral gaps*, J. Funct. Anal. 176 (2000), no. 2, 368–399.

[10] R. Bass, *Probabilistic techniques in analysis*, Springer-Verlag, New York, 1995.

[11] R.F. Bass, K. Burdzy, *A probabilistic proof of the boundary Harnack principle*, In E. Cinlar, K.L. Chung, and R.K. Getoor, editors, Seminar on Stochastic Processes, 1989, Birkhäuser, 1–16, Boston, 1990.

[12] R. Bass, M. Kassmann, *Harnack inequalities for non-local operators of variable order*, Trans. Amer. Math. Soc. 357 (2005), 837–850.

[13] R. Bass, D.A. Levin, *Harnack inequalities for jump processes*, Potential Anal. 17 (2002), 375–388.

[14] R. Bass, D.A. Levin, *Transition probabilities for symmetric jump processes*, Trans. Amer. Math. Soc. 354 (2002), 2933–2953.

[15] R. Bass, D. You, *A Fatou theorem for $\alpha$-harmonic functions*, Bull. Sciences Math. 127 (2003), 635–648.

[16] M. van den Berg, *On condensation in the free–boson gas and the spectrum of the Laplacian*, J. Statist. Phys. 31 (1983), 623–637.

[17] C. Berg, G. Forst, *Potential Theory on Locally Compact Abelian Groups*, Springer, Berlin 1975.

[18] J. Bertoin, *Lévy processes*, Cambridge Tracts in Mathematics 121, Cambridge 1996.

[19] J. Bertoin, *Regenerative embeddings of Markov sets*, Probab. Theory Rel. Fields 108 (1997), 559–571.

[20] Bertoin, J., Yor, M.: On subordinators, self-similar Markov processes and some factorizations of the exponential random variable. Elect. Comm. in Probab. 6 (2001), 95–106.

[21] N.H. Bingham, C.M. Goldie, J.L. Teugels, *Regular Variation*, Cambridge University Press, Cambridge 1987.

[22] J. Bliedtner, W. Hansen, *Potential Theory. An analytic and probabilistic approach to balayage*, Springer-Verlag, Berlin Heidelberg 1986.

[23] R.M. Blumenthal and R.K. Getoor, *Markov processes and potential theory*, Springer-Verlag, New York 1968.

[24] R.M. Blumenthal and R.K. Getoor, *The asymptotic distribution of the eigenvalues for a class of Markov operators* Pacific J. Math. 9 (1959), 399–408.

[25] R.M. Blumenthal, R.K. Getoor, D.B. Ray, *On the distribution of first hits for the symmetric stable processes*, Trans. Amer. Math. Soc. 99 (1961), 540–554.

[26] R.M. Blumenthal, R.K. Getoor, *Some theorems on stable processes*, Trans. Amer. Math. Soc. 95 (1960), 263–273.

[27] K. Bogdan, *The boundary Harnack principle for the fractional Laplacian*, Studia Math. 123 (1997), 43–80.

[28] K. Bogdan, *Representation of α-harmonic functions in Lipschitz domains*, Hiroshima J. Math. 29 (1999), 227–243.

[29] K. Bogdan, *Sharp estimates for the Green function in Lipschitz domains*, J. Math. Anal. Appl. 243 (2000), no. 2, 326–337.

[30] K. Bogdan, K. Burdzy, Z.-Q. Chen, *Censored stable processes*, Probab. Theory Related Fields 127 (2003), no. 1, 89–152.

[31] K. Bogdan and T. Byczkowski, *Probabilistic proof of the boundary Harnack principle for symmetric stable processes*, Potential Anal. 11 (1999), 135–156.

[32] K. Bogdan and T. Byczkowski, *Potential theory for the α-stable Schrödinger operator on on bounded Lipschitz domains*, Studia Math. 133 (1999), 53–92.

[33] K. Bogdan and T. Byczkowski, *Potential theory of Schrödinger operator based on fractional Laplacian*. Probab. Math. Statist. 20 (2000), no. 2, Acta Univ. Wratislav. No. 2256, 293–335.

[34] K. Bogdan and B. Dyda, *Relative Fatou theorem for harmonic and q-harmonic functions of rotation invariant stable process in smooth domains*, Studia Math. 157 (2003), 83–96.

[35] K. Bogdan and T. Jakubowski, *Probléme de Dirichlet pour les fonctions α-harmoniques.* (French) [Dirichlet problem for α-harmonic functions on conical domains] Ann. Math. Blaise Pascal 12 (2005), no. 2, 297–308.

[36] K. Bogdan and T. Jakubowski, *Estimates of heat kernel of fractional Laplacian perturbed by gradient operators*, Comm. Math. Phys. 271 (2007), no. 1, 179–198.

[37] K. Bogdan, T. Jakubowski and W. Hansen, *Time-dependent Schrödinger perturbations of transition densities*, Studia Math. 189 (2008), no. 3, 235–254.

[38] K. Bogdan, T. Kulczycki, M. Kwaśnicki, *Estimates and structure of α-harmonic functions*, Prob. Theory Related Fields 140 (2008), no. 3-4, 345–381.

[39] K. Bogdan, A. Stós, P. Sztonyk, *Harnack inequality for stable processes on d-sets*, Studia Math. 158 (2003), no. 2, 163–198.

[40] K. Bogdan, P. Sztonyk, *Estimates of the potential kernel and Harnack's inequality for the anisotropic fractional Laplacian*, Studia Math. 181 (2007), no. 2, 101–123.

[41] K. Bogdan, T. Żak, *On Kelvin transformation*, J. Theor. Prob. 19 (2006), no. 1, 89–120.

[42] M. Brelot, *On topologies and boundaries in potential theory*, Lecture Notes in Mathematics, Springer, Berlin, 1971.

[43] K. Burdzy, T. Kulczycki, *Stable processes have thorns*, Ann. Probab. 31 (2003), 170–194.

[44] T. Byczkowski, J. Małecki, M. Ryznar, *Bessel Potentials, Hitting Distributions and Green Functions*, Trans. Amer. Math. Soc. 361 (2009), 4871–4900.

[45] R. Carmona, W.C. Masters and B. Simon, *Relativistic Schrödinger operators: asymptotic behavior of the eigenfunctions*, J. Funct. Anal. 91 (1990), 117–142.

[46] Z.-Q. Chen, R. Durret and G. Ma, *Holomorphic diffusions and boundary behaviour of harmonic functions*, Ann. Prob. 25 (1997), 1103–1134.

[47] Z.-Q. Chen, P.J. Fitzsimmons, M. Takeda, J. Ying and T.-S. Zhang, *Absolute continuity of symmetric Markov processes*, Ann. Probab. 32 (2004), 2067–2098.

[48] Z.-Q. Chen, P. Kim, *Green function estimate for censored stable processes*, Probab. Theory Related Fields 124 (2002), no. 4, 595–610.

[49] Z.-Q. Chen, T. Kumagai, *Heat kernel estimates for stable-like processes on d-sets*, Stoch. Proc. Appl. 108 (2003), 27–62.

[50] Z.-Q. Chen and R. Song, *Martin boundary and integral representation for harmonic functions of symmetric stable processes*, J. Funct. Anal. 159 (1998), 267–294.

[51] Z.-Q. Chen and R. Song, *Estimates on Green functions and Poisson kernels for symmetric stable processes*, Math. Ann. 312 (1998), 465–501.

[52] Z.-Q. Chen and R. Song, *Conditional gauge theorem for non-local Feynman-Kac transforms*, Probab. Theory Related Fields 125 (2003), no. 1, 45–72.

[53] Z.-Q. Chen and R. Song, *Drift transforms and Green function estimates for discontinuous processes*, J. Funct. Anal. 201 (2003), 262–281.

[54] Z.-Q. Chen and R. Song, *Intrinsic ultracontractivity and Conditional Gauge for symmetric stable processes*, J. Funct. Anal. 150 (1997), no.1, 204–239.

[55] Z.-Q. Chen and R. Song, *Intrinsic ultracontractivity, conditional lifetimes and conditional gauge for symmetric stable processes on rough domains*, Illinois J. Math. 44 (2000), no. 1, 138–160.

[56] Z.-Q. Chen and R. Song, *Martin boundary and integral representation for harmonic functions of symmetric stable processes*, J. Funct. Anal. 159 (1998), 267–294.

[57] Z.-Q. Chen, R. Song, *Two-sided eigenvalue estimates for subordinate Brownian motion in bounded domains*, J. Funct. Anal. 226 (2005), 90–113.

[58] Z.-Q. Chen, R. Song, *Continuity of eigenvalues of subordinate processes in domains*, Math. Z. 252 (2006), 71–89.

[59] Z.-Q. Chen, R. Song, *Spectral properties of subordinate processes in domains*, in: Stochastic Analysis and Partial Differential Equations, 77–84, Contemp. Math. 429, AMS, Providence 2007.

[60] K.-L. Chung, *Lectures from Markov processes to Brownian motion*, Springer-Verlag, New York 1982.

[61] K.-L. Chung, K.M. Rao, *General gauge theorem for multiplicative functionals*, Trans. Amer. Math. Soc. 306 (1988), 819–836.

[62] K.-L. Chung, Z. Zhao, *From Brownian motion to Schrödinger's equation*, Springer-Verlag, New York, 1995.

[63] R. Courant, D. Hilbert, *Methods of mathematical physics*, vol. I, Interscience Publishers, Inc., New York, 1953.

[64] M. Cranston, E. Fabes and Z. Zhao, *Conditional gauge and potential theory for the Schrödinger operator*, Trans. Amer. Math. Soc. 307 (1988), 174–194.

[65] B. Dahlberg, *Estimates of harmonic measure*, Arch. Ration. Mech. Anal. 65 (1977), 275–288.

[66] E.B. Davies, Heat Kernels and Spectral Theory, Cambridge University Press, Cambridge 1989.

[67] E.B. Davies, B. Simon, *Ultracontractivity and the heat kernel for Schrödinger operators and Dirichlet Laplacians*, J. Funct. Anal. 59 (1984), 335–395.

[68] B. Davis, *On the spectral gap for fixed membranes*, Ark. Mat. 39 (2001), no. 1, 65–74.

[69] R.D. DeBlassie and P.J. Méndez-Hernández, *α–continuity properties of symmetric α-stable process*, Trans. Amer. Math. Soc. 359 (2007), no. 5, 2343–2359 (electronic).

[70] B. Dittmar, A.Y. Solynin, *The mixed Steklov eigenvalue problem and new extremal properties of the Grötzsch ring. (Russian)*, Zap. Nauchn. Sem. S.-Peterburg. Otdel. Mat. Inst. Steklov. (POMI) 270 (2000), Issled. po Linein. Oper. i Teor. Funkts. 28, 51–79, 365.

[71] J.L. Doob, *Classical Potential Theory and Its Probabilistic Counterpart*, Springer-Verlag, New York 1984.

[72] J.L. Doob, *A relativized Fatou theorem*, Proc. Nat. Acad. Sci. U.S.A. 45 (1959), 215–222.

[73] B. Dyda, T. Kulczycki *Spectral gap for stable process on convex double symmetric domains*, Potential Anal. 27 (2007), no. 2, 101–132.

[74] A. Erdèlyi, W. Magnus, F. Oberhettinger, F.G. Tricomi, *Tables of Integral Transforms*, Vol. 1., McGraw-Hill, New York 1954.

[75] P. Fatou, *Séries trigonométriques et series de Taylor*, Acta Math. 30 (1906), 335–400.

[76] B. Fuglede, *On the theory of potentials in locally compact spaces*, Acta Math. 103 (1960), 139–215.

[77] H. Geman, D.B. Madan, M. Yor, *Time changes for Lévy processes*, Math. Finance, 11 (2001), 79–96.

[78] R.K. Getoor, *First passage times for symmetric stable processes in space.* Trans. Amer. Math. Soc. 101 (1961), 75–90.

[79] R.K. Getoor, *Markov operators and their associated semi-groups*, Pacific J. Math. 9 (1959), 449–472.

[80] J. Glover, Z. Pop-Stojanovic, M. Rao, H. Šikić, R. Song, Z. Vondraček, *Harmonic functions of subordinate killed Brownian motions*, J. Funct. Anal., 215 (2004), 399–426.

[81] J. Glover, M. Rao, H. Šikić, R. Song, *Γ-potentials*, in: Classical and modern potential theory and applications (Chateau de Bonas, 1993), 217–232, Kluwer Acad. Publ., Dordrecht 1994.

[82] T. Grzywny, M. Ryznar, *Estimates of Green function for some perturbations of fractional Laplacian*, Illinois J. Math. 51 (2007), no. 4, 1409–1438

[83] T. Grzywny, M. Ryznar, *Two-sided optimal bounds for half-spaces for relativistic α-stable process*, Potential Anal. 28 (2008), no. 3, 201–239.

[84] W. Hansen, *Uniform boundary Harnack principle and generalized triangle property*, J. Funct. Anal. 226 (2005), no.2, 452–484.

[85] W. Hansen, *Global comparison of perturbed Green functions*, Math. Ann. 334(3) (2006), 643–678.

[86] J. Hawkes, *A lower Lipschitz condition for the stable subordinator*, Z. Wahrscheinlichkeitstheorie und Verw. Gebiete 17 (1971), 23–32.

[87] P. He, J. Ying, *Subordinating the killed process versus killing the subordinate process*, Proceedings Amer. Math. Soc. 135 (2007), 523–529.

[88] J. Hersch, L.E. Payne, *Extremal principles and isoperimetric inequalities for some mixed problems of Stekloff's type*, Z. Angew. Math. Phys. 19 (1968) 802–817.

[89] F. Hmissi, *Fonctions harmoniques pour les potentiels de Riesz sur la boule unite* (French) [*Harmonic functions for Riesz potentials on the unit ball*], Exposition. Math. 12 (1994), no. 3, 281–288.

[90] W. Hoh, *Pseudo differential operators generating Markov processes*, Habilitationsschrift, Universität Bielefeld 1998.

[91] F. Hubalek, A.E. Kyprianou, *Old and new examples of scale functions for spectrally negative Lévy processes*, preprint (2007).

[92] G.A. Hunt, *Markoff processes and potentials I, II, III*, Illinois J. Math. 1 (1957), 44–93; 1 (1957), 316–369; 2 (1957), 151–213.

[93] R.A. Hunt, R.L. Wheeden, *Positive harmonic functions on Lipschitz domains*, Trans. Amer. Math. Soc. 147 (1970), 507–527.

[94] R.A. Hunt, R.L. Wheeden, *On the boundary value of harmonic functions*, Trans. Amer. Math. Soc. 132 (1968), 307–322.

[95] N. Ikeda, S. Watanabe, *On some relations between the harmonic measure and the Lévy measure for a certain class of Markov processes*, Probab. Theory Related Fields 114 (1962), 207–227.

[96] N. Ikeda and S. Watanabe, *On some relations between the harmonic measure and the Lévy measure for a certain class of Markov processes*, J. Math. Kyoto Univ. 2 (1962), no. 1, 79–95.

[97] N. Jacob, *Pseudo differential operators and Markov processes. Vol. I. Fourier analysis and semigroups*, Imperial College Press, London, 2001.

[98] T. Jakubowski, *The estimates for the Green function in Lipschitz domains for the symmetric stable processes*, Probab. Math. Statist. 22 (2002), no. 2, Acta Univ. Wratislav. No. 2470, 419–441.

[99] D.S. Jerison, C.E. Kenig, *Boundary value problems on Lipschitz domains. Studies in partial differential equations*, MAA Stud. Math. 23 (1982), 1–68.

[100] D.S. Jerison, C.E. Kenig, *Boundary behavior of harmonic functions in non-tangentially accessible domains*, Advances in Math. 46 (1982), 80–147.

[101] P. Kim, *Relative Fatou's theorem for* $-(-\Delta)^{-\alpha/2}$ *harmonic functions in $\kappa$-fat set*, J. Funct. Anal. 234 (2006), no. 1, 70–105.

[102] P. Kim, Y.-R. Lee, *Generalized 3G theorem and application to relativistic stable process on non-smooth open sets*, J. Funct. Anal. 246 (2007), no. 1, 113–143.

[103] P. Kim, R. Song, *Potential theory of truncated stable processes*, Math. Z. 256 (2007), no. 1, 139–173.

[104] P. Kim, R. Song, *Boundary behavior of harmonic functions for truncated stable processes*, J. Theoret. Probab. 21 (2008), no. 2, 287–321.

[105] P. Kim, R. Song, Z. Vondraček, *Boundary Harnack principle for subordinate Brownian motion*, Stochastic Process. Appl. 119 (2009), 1601–1631.

[106] L.B. Klebanov, G.M. Maniya, I.A. Melamed, *A problem of V.M. Zolotarev and analogues of infinitely divisible and stable distributions in a scheme for summation of a random number of random variables*, Theory Probab. Appl. 29 (1984), 791–794.

[107] P. Koosis, *Introduction to $H_p$ spaces*, LMS Lecture Note Series 40, Cambridge University Press, 1980.

[108] V. Kozlov, N. Kuznetsov, O. Motygin, *On the two-dimensional sloshing problem* Proc. Roy. Soc. London A. 460 (2004), no. 2049, 2587–2603.

[109] T. Kulczycki, *Properties of Green function of symmetric stable processes*, Probab. Math. Statist. 17 (1997), no. 2, Acta Univ. Wratislav. no. 2029, 339–364.

[110] T. Kulczycki, *Intrinsic ultracontractivity for symmetric stable processes*, Bull. Polish Acad. Sci. Math. 46 (1998), no. 3, 325–334.

[111] T. Kulczycki, *Exit time and Green function of cone for symmetric stable processes*, Probab. Math. Statist. 19 (1999), no. 2, Acta Univ. Wratislav. No. 2198, 337–374.

[112] T. Kulczycki, B. Siudeja, *Intrinsic ultracontractivity of Feynman-Kac semigroup for relativistic stable processes*, Trans. Amer. Math. Soc. 358, no. 11 (2006), 5025–5057.

[113] H. Kunita, T. Watanabe, *Markov processes and Martin boundaries I*, Illinois J. Math. 9 (1965), 485–526.

[114] I.A. Kunugui, *Étude sur la théorie du potentiel généralisé*, Osaka Math. J. 2 (1950), 63–103.

[115] N. Kuznetsov, O. Motygin, *Sloshing problem in a half-plane covered by a dock with two gaps: monotonicity and asymptotics of eigenvalues* Comptes Rendus de l'Academie des Sciences: Serie II b, Mechanique, Physique, Astronomie 329 (11) (2001), 791–796.

[116] M. Kwaśnicki, *Spectral gap estimate for stable processes on arbitrary bounded open sets*, Probab. Math. Statist. 28(1) (2008), 163–167.

[117] N.S. Landkof, *Foundations of modern potential theory*, Springer-Verlag, New York 1972.

[118] E.H. Lieb, *The stability of matter: from atoms to stars*, Bull. Amer. Math. Soc. 22 (1990), 1–49.

[119] J. Ling, *A lower bound for the gap between the first two eigenvalues of Schrödinger operators on convex domains in $S^n$ or $R^n$*, Michigan Math. J. 40 (1993), no. 2, 259–270.

[120] J.E. Littlewood, *On a theorem of Fatou*, Proc. London Math.Soc. 2 (1927), 172–176.

[121] D.B. Madan, P. Carr, E. Chang, *The variance gamma process and option pricing*, European Finance Review 2 (1998), 79–105.

[122] H. Matsumoto, L. Nguyen, M. Yor, *Subordinators related to the exponential functionals of Brownian bridges and explicit formulae for the semigroups of hyperbolic Brownian motions*, in: Stochastic processes and related topics (Siegmundsburg, 2000), Stochastics Monogr. Vol. 12, 213–235, Taylor & Francis, London 2002.

[123] K. Michalik, *Sharp estimates of the Green function, the Poisson kernel and the Martin kernel of cones for symmetric stable processes*, Hiroshima Math. J. 36 (2006), no. 1, 1–21.

[124] K. Michalik, M. Ryznar, *Relative Fatou theorem for α-harmonic functions in Lipschitz domains*, Illinois J. Math. 48 (2004), no. 3, 977–998.

[125] K. Michalik, K. Samotij, *Martin representation for α-harmonic functions*, Probab. Math. Statist. 20 (2000), 75–91.

[126] Y. Nakamura, *Classes of operator monotone functions and Stieltjes functions*, in: The Gohberg Anniversary Collection, Vol. II: Topics in Analysis and Operator Theory, Operator Theory: Advances and Applications, Vol. 41, H. Dym et al., Eds, 395–404, Birkhäuser, Basel 1989.

[127] A.N. Pillai, *On Mittag-Leffler functions and related distributions*, Ann. Inst. Statist. Math., 42 (1990), 157–161.

[128] R. Pinsky, *Positive harmonic functions and diffusion*, Cambridge Studies in Advanced Mathematics, 45. Cambridge University Press, Cambridge 1995.

[129] S. Port, *Hitting times and potentials for recurrent stable processes*, J. Analyse Math. 20 (1967), 371–395.

[130] S.C. Port, J.C. Stone, *Infinitely divisible process and their potential theory I, II*, Ann. Inst. Fourier, 21(1971), no. 2, 157–275 and 21 (1971), no. 4, 179–265.

[131] W.E. Pruitt, *The growth of random walks and Lévy processes*, Ann. Probab., 9 (1981), 948–956.

[132] M. Rao, R. Song, Z. Vondraček, *Green function estimates and Harnack inequality for subordinate Brownian motions*, Potential Anal. 25 (2006), 1–27.

[133] M. Riesz, *Intégrales de Riemann-Liouville et potentiels*. Acta Sci. Math. Szeged, 1938.

[134] M. Ryznar, *Estimates of Green function for relativistic α-stable processes*, Potential Anal. 17 (2002), 1–23.

[135] K. Sato, *Lévy processes and infinitely divisible distributions*, Cambridge Univ. Press, Cambridge 1999.

[136] R.L. Schilling, *Subordination in the sense of Bochner and a related functional calculus*, J. Austral. Math. Soc., Ser. A, 64 (1998), 368–396.

[137] R.L. Schilling, *Growth and Hölder conditions for the sample paths of Feller processes*, Probab. Theory Related Fields 112 (1998), 565–611.

[138] R.L. Schilling, R. Song and Z. Vondraček, *Bernstein Functions - Theory and Applications*, forthcoming book.

[139] H. Šikić, R. Song, Z. Vondraček, *Potential theory of geometric stable processes*, Probab. Theory Related Fields 135 (2006), 547–575.

[140] B. Simon, *Schrödinger semigroups*, Bull. Amer. Math. Soc. 7 (1982), 447–526.

[141] I.M. Singer, B. Wong, S.-T. Yau, S.S.-T. Yau, *An estimate of the gap of the first two eigenvalues in the Schrödinger operator*, Ann. Scuola Norm. Sup. Pisa Cl. Sci. (4) 12(2) (1985), 319–333.

[142] A.V. Skorohod, *Asymptotic formulas for stable distribution laws*, Select. Transl. Math. Statist. and Probability, Vol. 1., 157–161, Inst. Math. Statist. and Amer. Math. Soc., Providence, R.I., 1961.

[143] R. Song, *Sharp bounds on the density, Green function and jumping function of subordinate killed Brownian Motion*, Probab. Theory Related Fields 128 (2004), 606–628.

[144] R. Song, Z. Vondraček, *Potential theory of subordinate killed Brownian motion in a domain*, Probab. Theory Related Fields 125 (2003), 578–592.

[145] R. Song, Z. Vondraček, *Harnack inequalities for some classes of Markov processes*, Math. Z. 246 (2004), 177–202.

[146] R. Song, Z. Vondraček, *Sharp bounds for Green functions and jumping functions of subordinate killed Brownian motions in bounded $C^{1,1}$ domains*, Elect. Comm. in Probab. 9 (2004), 96–105.

[147] R. Song, Z. Vondraček, *Harnack inequality for some discontinuous Markov processes with a diffusion part*, Glasnik Matematički 40 (2005), 177–187.

[148] R. Song, Z. Vondraček, *Potential theory of special subordinators and subordinate killed stable processes*, J. Theoret. Probab. 19 (2006), 817–847.

[149] R. Song, Z. Vondraček, *On the relationship between subordinate killed and killed subordinate process*, Electron. Commun. Probab. 13 (2008), 325–336.

[150] R. Song, J.-M. Wu, *Boundary Harnack principle for symmetric stable processes*, J. Funct. Anal. 168 (1999), 403–427.

[151] D.W. Stroock, *Markov processes from K. Itô's perspective*, Ann. of Math. Stud. 155, Princeton Univ. Press, Princeton, NJ, 2003.

[152] P. Sztonyk, *On harmonic measure for Lévy processes*, Probab. Math. Statist. 20 (2000), 383–390.

[153] J.-M. Wu, *Comparisons of kernel functions, boundary Harnack principle and relative Fatou theorem on Lipschitz domains*, Ann. Inst. Fourier Grenoble 28 (1978), 147–167.

[154] J.-M. Wu, *Harmonic measures for symmetric stable processes*, Studia Math. 149 (2002), no. 3, 281–293.

[155] J.-M. Wu, *Symmetric stable processes stay in thick sets*, Ann. Probab. 32 (2004), 315–336.

[156] K. Yosida, *Functional analysis*, Springer, 1995.

[157] Q. Yu and J.Q. Zhong, *Lower bounds of the gap between the first and second eigenvalues of the Schrödinger operator*, Trans. Amer. Math. Soc. 294 (1986), 341–349.

[158] Q.S. Zhang, *The boundary behavior of heat kernels of Dirichlet Laplacians*, J. Diff. Equations 182 (2002), 416–430.

# Index

# Lecture Notes in Mathematics

For information about earlier volumes
please contact your bookseller or Springer
LNM Online archive: springerlink.com

Vol. 1837: S. Tavaré, O. Zeitouni, Lectures on Probability Theory and Statistics. Ecole d'Eté de Probabilités de Saint-Flour XXXI-2001. Editor: J. Picard (2004)

Vol. 1838: A.J. Ganesh, N.W. O'Connell, D.J. Wischik, Big Queues. XII, 254 p, 2004.

Vol. 1839: R. Gohm, Noncommutative Stationary Processes. VIII, 170 p, 2004.

Vol. 1840: B. Tsirelson, W. Werner, Lectures on Probability Theory and Statistics. Ecole d'Eté de Probabilités de Saint-Flour XXXII-2002. Editor: J. Picard (2004)

Vol. 1841: W. Reichel, Uniqueness Theorems for Variational Problems by the Method of Transformation Groups (2004)

Vol. 1842: T. Johnsen, A. L. Knutsen, $K_3$ Projective Models in Scrolls (2004)

Vol. 1843: B. Jefferies, Spectral Properties of Noncommuting Operators (2004)

Vol. 1844: K.F. Siburg, The Principle of Least Action in Geometry and Dynamics (2004)

Vol. 1845: Min Ho Lee, Mixed Automorphic Forms, Torus Bundles, and Jacobi Forms (2004)

Vol. 1846: H. Ammari, H. Kang, Reconstruction of Small Inhomogeneities from Boundary Measurements (2004)

Vol. 1847: T.R. Bielecki, T. Björk, M. Jeanblanc, M. Rutkowski, J.A. Scheinkman, W. Xiong, Paris-Princeton Lectures on Mathematical Finance 2003 (2004)

Vol. 1848: M. Abate, J. E. Fornaess, X. Huang, J. P. Rosay, A. Tumanov, Real Methods in Complex and CR Geometry, Martina Franca, Italy 2002. Editors: D. Zaitsev, G. Zampieri (2004)

Vol. 1849: Martin L. Brown, Heegner Modules and Elliptic Curves (2004)

Vol. 1850: V. D. Milman, G. Schechtman (Eds.), Geometric Aspects of Functional Analysis. Israel Seminar 2002-2003 (2004)

Vol. 1851: O. Catoni, Statistical Learning Theory and Stochastic Optimization (2004)

Vol. 1852: A.S. Kechris, B.D. Miller, Topics in Orbit Equivalence (2004)

Vol. 1853: Ch. Favre, M. Jonsson, The Valuative Tree (2004)

Vol. 1854: O. Saeki, Topology of Singular Fibers of Differential Maps (2004)

Vol. 1855: G. Da Prato, P.C. Kunstmann, I. Lasiecka, A. Lunardi, R. Schnaubelt, L. Weis, Functional Analytic Methods for Evolution Equations. Editors: M. Iannelli, R. Nagel, S. Piazzera (2004)

Vol. 1856: K. Back, T.R. Bielecki, C. Hipp, S. Peng, W. Schachermayer, Stochastic Methods in Finance, Bressanone/Brixen, Italy, 2003. Editors: M. Fritelli, W. Runggaldier (2004)

Vol. 1857: M. Émery, M. Ledoux, M. Yor (Eds.), Séminaire de Probabilités XXXVIII (2005)

Vol. 1858: A.S. Cherny, H.-J. Engelbert, Singular Stochastic Differential Equations (2005)

Vol. 1859: E. Letellier, Fourier Transforms of Invariant Functions on Finite Reductive Lie Algebras (2005)

Vol. 1860: A. Borisyuk, G.B. Ermentrout, A. Friedman, D. Terman, Tutorials in Mathematical Biosciences I. Mathematical Neurosciences (2005)

Vol. 1861: G. Benettin, J. Henrard, S. Kuksin, Hamiltonian Dynamics – Theory and Applications, Cetraro, Italy, 1999. Editor: A. Giorgilli (2005)

Vol. 1862: B. Helffer, F. Nier, Hypoelliptic Estimates and Spectral Theory for Fokker-Planck Operators and Witten Laplacians (2005)

Vol. 1863: H. Führ, Abstract Harmonic Analysis of Continuous Wavelet Transforms (2005)

Vol. 1864: K. Efstathiou, Metamorphoses of Hamiltonian Systems with Symmetries (2005)

Vol. 1865: D. Applebaum, B.V. R. Bhat, J. Kustermans, J. M. Lindsay, Quantum Independent Increment Processes I. From Classical Probability to Quantum Stochastic Calculus. Editors: M. Schürmann, U. Franz (2005)

Vol. 1866: O.E. Barndorff-Nielsen, U. Franz, R. Gohm, B. Kümmerer, S. Thorbjønsen, Quantum Independent Increment Processes II. Structure of Quantum Lévy Processes, Classical Probability, and Physics. Editors: M. Schürmann, U. Franz, (2005)

Vol. 1867: J. Sneyd (Ed.), Tutorials in Mathematical Biosciences II. Mathematical Modeling of Calcium Dynamics and Signal Transduction. (2005)

Vol. 1868: J. Jorgenson, S. Lang, $Pos_n(R)$ and Eisenstein Series. (2005)

Vol. 1869: A. Dembo, T. Funaki, Lectures on Probability Theory and Statistics. Ecole d'Eté de Probabilités de Saint-Flour XXXIII-2003. Editor: J. Picard (2005)

Vol. 1870: V.I. Gurariy, W. Lusky, Geometry of Müntz Spaces and Related Questions. (2005)

Vol. 1871: P. Constantin, G. Gallavotti, A.V. Kazhikhov, Y. Meyer, S. Ukai, Mathematical Foundation of Turbulent Viscous Flows, Martina Franca, Italy, 2003. Editors: M. Cannone, T. Miyakawa (2006)

Vol. 1872: A. Friedman (Ed.), Tutorials in Mathematical Biosciences III. Cell Cycle, Proliferation, and Cancer (2006)

Vol. 1873: R. Mansuy, M. Yor, Random Times and Enlargements of Filtrations in a Brownian Setting (2006)

Vol. 1874: M. Yor, M. Émery (Eds.), In Memoriam Paul-André Meyer - Séminaire de Probabilités XXXIX (2006)

Vol. 1875: J. Pitman, Combinatorial Stochastic Processes. Ecole d'Eté de Probabilités de Saint-Flour XXXII-2002. Editor: J. Picard (2006)

Vol. 1876: H. Herrlich, Axiom of Choice (2006)

Vol. 1877: J. Steuding, Value Distributions of $L$-Functions (2007)

Vol. 1878: R. Cerf, The Wulff Crystal in Ising and Percolation Models, Ecole d'Eté de Probabilités de Saint-Flour XXXIV-2004. Editor: Jean Picard (2006)

Vol. 1879: G. Slade, The Lace Expansion and its Applications, Ecole d'Eté de Probabilités de Saint-Flour XXXIV-2004. Editor: Jean Picard (2006)

Vol. 1880: S. Attal, A. Joye, C.-A. Pillet, Open Quantum Systems I, The Hamiltonian Approach (2006)

Vol. 1881: S. Attal, A. Joye, C.-A. Pillet, Open Quantum Systems II, The Markovian Approach (2006)

Vol. 1882: S. Attal, A. Joye, C.-A. Pillet, Open Quantum Systems III, Recent Developments (2006)

Vol. 1883: W. Van Assche, F. Marcellàn (Eds.), Orthogonal Polynomials and Special Functions, Computation and Application (2006)

Vol. 1884: N. Hayashi, E.I. Kaikina, P.I. Naumkin, I.A. Shishmarev, Asymptotics for Dissipative Nonlinear Equations (2006)

Vol. 1885: A. Telcs, The Art of Random Walks (2006)

Vol. 1886: S. Takamura, Splitting Deformations of Degenerations of Complex Curves (2006)

Vol. 1887: K. Habermann, L. Habermann, Introduction to Symplectic Dirac Operators (2006)

Vol. 1888: J. van der Hoeven, Transseries and Real Differential Algebra (2006)

Vol. 1889: G. Osipenko, Dynamical Systems, Graphs, and Algorithms (2006)

Vol. 1890: M. Bunge, J. Funk, Singular Coverings of Toposes (2006)

Vol. 1891: J.B. Friedlander, D.R. Heath-Brown, H. Iwaniec, J. Kaczorowski, Analytic Number Theory, Cetraro, Italy, 2002. Editors: A. Perelli, C. Viola (2006)

Vol. 1892: A. Baddeley, I. Bárány, R. Schneider, W. Weil, Stochastic Geometry, Martina Franca, Italy, 2004. Editor: W. Weil (2007)

Vol. 1893: H. Hanßmann, Local and Semi-Local Bifurcations in Hamiltonian Dynamical Systems, Results and Examples (2007)

Vol. 1894: C.W. Groetsch, Stable Approximate Evaluation of Unbounded Operators (2007)

Vol. 1895: L. Molnár, Selected Preserver Problems on Algebraic Structures of Linear Operators and on Function Spaces (2007)

Vol. 1896: P. Massart, Concentration Inequalities and Model Selection, Ecole d'Été de Probabilités de Saint-Flour XXXIII-2003. Editor: J. Picard (2007)

Vol. 1897: R. Doney, Fluctuation Theory for Lévy Processes, Ecole d'Été de Probabilités de Saint-Flour XXXV-2005. Editor: J. Picard (2007)

Vol. 1898: H.R. Beyer, Beyond Partial Differential Equations, On linear and Quasi-Linear Abstract Hyperbolic Evolution Equations (2007)

Vol. 1899: Séminaire de Probabilités XL. Editors: C. Donati-Martin, M. Émery, A. Rouault, C. Stricker (2007)

Vol. 1900: E. Bolthausen, A. Bovier (Eds.), Spin Glasses (2007)

Vol. 1901: O. Wittenberg, Intersections de deux quadriques et pinceaux de courbes de genre 1, Intersections of Two Quadrics and Pencils of Curves of Genus 1 (2007)

Vol. 1902: A. Isaev, Lectures on the Automorphism Groups of Kobayashi-Hyperbolic Manifolds (2007)

Vol. 1903: G. Kresin, V. Maz'ya, Sharp Real-Part Theorems (2007)

Vol. 1904: P. Giesl, Construction of Global Lyapunov Functions Using Radial Basis Functions (2007)

Vol. 1905: C. Prévôt, M. Röckner, A Concise Course on Stochastic Partial Differential Equations (2007)

Vol. 1906: T. Schuster, The Method of Approximate Inverse: Theory and Applications (2007)

Vol. 1907: M. Rasmussen, Attractivity and Bifurcation for Nonautonomous Dynamical Systems (2007)

Vol. 1908: T.J. Lyons, M. Caruana, T. Lévy, Differential Equations Driven by Rough Paths, Ecole d'Été de Probabilités de Saint-Flour XXXIV-2004 (2007)

Vol. 1909: H. Akiyoshi, M. Sakuma, M. Wada, Y. Yamashita, Punctured Torus Groups and 2-Bridge Knot Groups (I) (2007)

Vol. 1910: V.D. Milman, G. Schechtman (Eds.), Geometric Aspects of Functional Analysis. Israel Seminar 2004-2005 (2007)

Vol. 1911: A. Bressan, D. Serre, M. Williams, K. Zumbrun, Hyperbolic Systems of Balance Laws. Cetraro, Italy 2003. Editor: P. Marcati (2007)

Vol. 1912: V. Berinde, Iterative Approximation of Fixed Points (2007)

Vol. 1913: J.E. Marsden, G. Misiołek, J.-P. Ortega, M. Perlmutter, T.S. Ratiu, Hamiltonian Reduction by Stages (2007)

Vol. 1914: G. Kutyniok, Affine Density in Wavelet Analysis (2007)

Vol. 1915: T. Bıyıkoğlu, J. Leydold, P.F. Stadler, Laplacian Eigenvectors of Graphs. Perron-Frobenius and Faber-Krahn Type Theorems (2007)

Vol. 1916: C. Villani, F. Rezakhanlou, Entropy Methods for the Boltzmann Equation. Editors: F. Golse, S. Olla (2008)

Vol. 1917: I. Veselić, Existence and Regularity Properties of the Integrated Density of States of Random Schrödinger (2008)

Vol. 1918: B. Roberts, R. Schmidt, Local Newforms for GSp(4) (2007)

Vol. 1919: R.A. Carmona, I. Ekeland, A. Kohatsu-Higa, J.-M. Lasry, P.-L. Lions, H. Pham, E. Taflin, Paris-Princeton Lectures on Mathematical Finance 2004. Editors: R.A. Carmona, E. Çinlar, I. Ekeland, E. Jouini, J.A. Scheinkman, N. Touzi (2007)

Vol. 1920: S.N. Evans, Probability and Real Trees. Ecole d'Été de Probabilités de Saint-Flour XXXV-2005 (2008)

Vol. 1921: J.P. Tian, Evolution Algebras and their Applications (2008)

Vol. 1922: A. Friedman (Ed.), Tutorials in Mathematical BioSciences IV. Evolution and Ecology (2008)

Vol. 1923: J.P.N. Bishwal, Parameter Estimation in Stochastic Differential Equations (2008)

Vol. 1924: M. Wilson, Littlewood-Paley Theory and Exponential-Square Integrability (2008)

Vol. 1925: M. du Sautoy, L. Woodward, Zeta Functions of Groups and Rings (2008)

Vol. 1926: L. Barreira, V. Claudia, Stability of Nonautonomous Differential Equations (2008)

Vol. 1927: L. Ambrosio, L. Caffarelli, M.G. Crandall, L.C. Evans, N. Fusco, Calculus of Variations and Non-Linear Partial Differential Equations. Cetraro, Italy 2005. Editors: B. Dacorogna, P. Marcellini (2008)

Vol. 1928: J. Jonsson, Simplicial Complexes of Graphs (2008)

Vol. 1929: Y. Mishura, Stochastic Calculus for Fractional Brownian Motion and Related Processes (2008)

Vol. 1930: J.M. Urbano, The Method of Intrinsic Scaling. A Systematic Approach to Regularity for Degenerate and Singular PDEs (2008)

Vol. 1931: M. Cowling, E. Frenkel, M. Kashiwara, A. Valette, D.A. Vogan, Jr., N.R. Wallach, Representation Theory and Complex Analysis. Venice, Italy 2004. Editors: E.C. Tarabusi, A. D'Agnolo, M. Picardello (2008)

Vol. 1932: A.A. Agrachev, A.S. Morse, E.D. Sontag, H.J. Sussmann, V.I. Utkin, Nonlinear and Optimal Control Theory. Cetraro, Italy 2004. Editors: P. Nistri, G. Stefani (2008)

Vol. 1933: M. Petkovic, Point Estimation of Root Finding Methods (2008)

Vol. 1934: C. Donati-Martin, M. Émery, A. Rouault, C. Stricker (Eds.), Séminaire de Probabilités XLI (2008)

Vol. 1935: A. Unterberger, Alternative Pseudodifferential Analysis (2008)

Vol. 1936: P. Magal, S. Ruan (Eds.), Structured Population Models in Biology and Epidemiology (2008)

Vol. 1937: G. Capriz, P. Giovine, P.M. Mariano (Eds.), Mathematical Models of Granular Matter (2008)

Vol. 1938: D. Auroux, F. Catanese, M. Manetti, P. Seidel, B. Siebert, I. Smith, G. Tian, Symplectic 4-Manifolds and Algebraic Surfaces. Cetraro, Italy 2003. Editors: F. Catanese, G. Tian (2008)

Vol. 1939: D. Boffi, F. Brezzi, L. Demkowicz, R.G. Durán, R.S. Falk, M. Fortin, Mixed Finite Elements,

Compatibility Conditions, and Applications. Cetraro, Italy 2006. Editors: D. Boffi, L. Gastaldi (2008)

Vol. 1940: J. Banasiak, V. Capasso, M.A.J. Chaplain, M. Lachowicz, J. Miękisz, Multiscale Problems in the Life Sciences. From Microscopic to Macroscopic. Będlewo, Poland 2006. Editors: V. Capasso, M. Lachowicz (2008)

Vol. 1941: S.M.J. Haran, Arithmetical Investigations. Representation Theory, Orthogonal Polynomials, and Quantum Interpolations (2008)

Vol. 1942: S. Albeverio, F. Flandoli, Y.G. Sinai, SPDE in Hydrodynamic. Recent Progress and Prospects. Cetraro, Italy 2005. Editors: G. Da Prato, M. Röckner (2008)

Vol. 1943: L.L. Bonilla (Ed.), Inverse Problems and Imaging. Martina Franca, Italy 2002 (2008)

Vol. 1944: A. Di Bartolo, G. Falcone, P. Plaumann, K. Strambach, Algebraic Groups and Lie Groups with Few Factors (2008)

Vol. 1945: F. Brauer, P. van den Driessche, J. Wu (Eds.), Mathematical Epidemiology (2008)

Vol. 1946: G. Allaire, A. Arnold, P. Degond, T.Y. Hou, Quantum Transport. Modelling, Analysis and Asymptotics. Cetraro, Italy 2006. Editors: N.B. Abdallah, G. Frosali (2008)

Vol. 1947: D. Abramovich, M. Mariño, M. Thaddeus, R. Vakil, Enumerative Invariants in Algebraic Geometry and String Theory. Cetraro, Italy 2005. Editors: K. Behrend, M. Manetti (2008)

Vol. 1948: F. Cao, J-L. Lisani, J-M. Morel, P. Musé, F. Sur, A Theory of Shape Identification (2008)

Vol. 1949: H.G. Feichtinger, B. Helffer, M.P. Lamoureux, N. Lerner, J. Toft, Pseudo-Differential Operators. Quantization and Signals. Cetraro, Italy 2006. Editors: L. Rodino, M.W. Wong (2008)

Vol. 1950: M. Bramson, Stability of Queueing Networks, Ecole d'Eté de Probabilités de Saint-Flour XXXVI-2006 (2008)

Vol. 1951: A. Moltó, J. Orihuela, S. Troyanski, M. Valdivia, A Non Linear Transfer Technique for Renorming (2009)

Vol. 1952: R. Mikhailov, I.B.S. Passi, Lower Central and Dimension Series of Groups (2009)

Vol. 1953: K. Arwini, C.T.J. Dodson, Information Geometry (2008)

Vol. 1954: P. Biane, L. Bouten, F. Cipriani, N. Konno, N. Privault, Q. Xu, Quantum Potential Theory. Editors: U. Franz, M. Schuermann (2008)

Vol. 1955: M. Bernot, V. Caselles, J.-M. Morel, Optimal Transportation Networks (2008)

Vol. 1956: C.H. Chu, Matrix Convolution Operators on Groups (2008)

Vol. 1957: A. Guionnet, On Random Matrices: Macroscopic Asymptotics, Ecole d'Eté de Probabilités de Saint-Flour XXXVI-2006 (2009)

Vol. 1958: M.C. Olsson, Compactifying Moduli Spaces for Abelian Varieties (2008)

Vol. 1959: Y. Nakkajima, A. Shiho, Weight Filtrations on Log Crystalline Cohomologies of Families of Open Smooth Varieties (2008)

Vol. 1960: J. Lipman, M. Hashimoto, Foundations of Grothendieck Duality for Diagrams of Schemes (2009)

Vol. 1961: G. Buttazzo, A. Pratelli, S. Solimini, E. Stepanov, Optimal Urban Networks via Mass Transportation (2009)

Vol. 1962: R. Dalang, D. Khoshnevisan, C. Mueller, D. Nualart, Y. Xiao, A Minicourse on Stochastic Partial Differential Equations (2009)

Vol. 1963: W. Siegert, Local Lyapunov Exponents (2009)

Vol. 1964: W. Roth, Operator-valued Measures and Integrals for Cone-valued Functions and Integrals for Cone-valued Functions (2009)

Vol. 1965: C. Chidume, Geometric Properties of Banach Spaces and Nonlinear Iterations (2009)

Vol. 1966: D. Deng, Y. Han, Harmonic Analysis on Spaces of Homogeneous Type (2009)

Vol. 1967: B. Fresse, Modules over Operads and Functors (2009)

Vol. 1968: R. Weissauer, Endoscopy for GSP(4) and the Cohomology of Siegel Modular Threefolds (2009)

Vol. 1969: B. Roynette, M. Yor, Penalising Brownian Paths (2009)

Vol. 1970: M. Biskup, A. Bovier, F. den Hollander, D. Ioffe, F. Martinelli, K. Netočný, F. Toninelli, Methods of Contemporary Mathematical Statistical Physics. Editor: R. Kotecký (2009)

Vol. 1971: L. Saint-Raymond, Hydrodynamic Limits of the Boltzmann Equation (2009)

Vol. 1972: T. Mochizuki, Donaldson Type Invariants for Algebraic Surfaces (2009)

Vol. 1973: M.A. Berger, L.H. Kauffmann, B. Khesin, H.K. Moffatt, R.L. Ricca, De W. Sumners, Lectures on Topological Fluid Mechanics. Cetraro, Italy 2001. Editor: R.L. Ricca (2009)

Vol. 1974: F. den Hollander, Random Polymers: École d'Été de Probabilités de Saint-Flour XXXVII – 2007 (2009)

Vol. 1975: J.C. Rohde, Cyclic Coverings, Calabi-Yau Manifolds and Complex Multiplication (2009)

Vol. 1976: N. Ginoux, The Dirac Spectrum (2009)

Vol. 1977: M.J. Gursky, E. Lanconelli, A. Malchiodi, G. Tarantello, X.-J. Wang, P.C. Yang, Geometric Analysis and PDEs. Cetraro, Italy 2001. Editors: A. Ambrosetti, S.-Y.A. Chang, A. Malchiodi (2009)

Vol. 1978: M. Qian, J.-S. Xie, S. Zhu, Smooth Ergodic Theory for Endomorphisms (2009)

Vol. 1979: C. Donati-Martin, M. Émery, A. Rouault, C. Stricker (Eds.), Séminaire de Probablitiés XLII (2009)

Vol. 1980: P. Graczyk, A. Stos (Eds.), Potential Analysis of Stable Processes and its Extensions (2009)

# Recent Reprints and New Editions

Vol. 1702: J. Ma, J. Yong, Forward-Backward Stochastic Differential Equations and their Applications. 1999 – Corr. 3rd printing (2007)

Vol. 830: J.A. Green, Polynomial Representations of $GL_n$, with an Appendix on Schensted Correspondence and Littelman Paths by K. Erdmann, J.A. Green and M. Schoker 1980 – 2nd corr. and augmented edition (2007)

Vol. 1693: S. Simons, From Hahn-Banach to Monotonicity (Minimax and Monotonicity 1998) – 2nd exp. edition (2008)

Vol. 470: R.E. Bowen, Equilibrium States and the Ergodic Theory of Anosov Diffeomorphisms. With a preface by D. Ruelle. Edited by J.-R. Chazottes. 1975 – 2nd rev. edition (2008)

Vol. 523: S.A. Albeverio, R.J. Høegh-Krohn, S. Mazzucchi, Mathematical Theory of Feynman Path Integral. 1976 – 2nd corr. and enlarged edition (2008)

Vol. 1764: A. Cannas da Silva, Lectures on Symplectic Geometry 2001 – Corr. 2nd printing (2008)

# LECTURE NOTES IN MATHEMATICS

Springer

Edited by J.-M. Morel, F. Takens, B. Teissier, P.K. Maini

**Editorial Policy** (for Multi-Author Publications: Summer Schools/Intensive Courses)

1. Lecture Notes aim to report new developments in all areas of mathematics and their applications - quickly, informally and at a high level. Mathematical texts analysing new developments in modelling and numerical simulation are welcome. Manuscripts should be reasonably self-contained and rounded off. Thus they may, and often will, present not only results of the author but also related work by other people. They should provide sufficient motivation, examples and applications. There should also be an introduction making the text comprehensible to a wider audience. This clearly distinguishes Lecture Notes from journal articles or technical reports which normally are very concise. Articles intended for a journal but too long to be accepted by most journals, usually do not have this "lecture notes" character.

2. In general SUMMER SCHOOLS and other similar INTENSIVE COURSES are held to present mathematical topics that are close to the frontiers of recent research to an audience at the beginning or intermediate graduate level, who may want to continue with this area of work, for a thesis or later. This makes demands on the didactic aspects of the presentation. Because the subjects of such schools are advanced, there often exists no textbook, and so ideally, the publication resulting from such a school could be a first approximation to such a textbook. Usually several authors are involved in the writing, so it is not always simple to obtain a unified approach to the presentation.

   For prospective publication in LNM, the resulting manuscript should not be just a collection of course notes, each of which has been developed by an individual author with little or no co-ordination with the others, and with little or no common concept. The subject matter should dictate the structure of the book, and the authorship of each part or chapter should take secondary importance. Of course the choice of authors is crucial to the quality of the material at the school and in the book, and the intention here is not to belittle their impact, but simply to say that the book should be planned to be written by these authors jointly, and not just assembled as a result of what these authors happen to submit.

   This represents considerable preparatory work (as it is imperative to ensure that the authors know these criteria before they invest work on a manuscript), and also considerable editing work afterwards, to get the book into final shape. Still it is the form that holds the most promise of a successful book that will be used by its intended audience, rather than yet another volume of proceedings for the library shelf.

3. Manuscripts should be submitted either online at www.editorialmanager.com/lnm/ to Springer's mathematics editorial, or to one of the series editors. Volume editors are expected to arrange for the refereeing, to the usual scientific standards, of the individual contributions. If the resulting reports can be forwarded to us (series editors or Springer) this is very helpful. If no reports are forwarded or if other questions remain unclear in respect of homogeneity etc, the series editors may wish to consult external referees for an overall evaluation of the volume. A final decision to publish can be made only on the basis of the complete manuscript; however a preliminary decision can be based on a pre-final or incomplete manuscript. The strict minimum amount of material that will be considered should include a detailed outline describing the planned contents of each chapter.

   Volume editors and authors should be aware that incomplete or insufficiently close to final manuscripts almost always result in longer evaluation times. They should also be aware that parallel submission of their manuscript to another publisher while under consideration for LNM will in general lead to immediate rejection.

4. Manuscripts should in general be submitted in English. Final manuscripts should contain at least 100 pages of mathematical text and should always include
   - a general table of contents;
   - an informative introduction, with adequate motivation and perhaps some historical remarks: it should be accessible to a reader not intimately familiar with the topic treated;
   - a global subject index: as a rule this is genuinely helpful for the reader.

   Lecture Notes volumes are, as a rule, printed digitally from the authors' files. We strongly recommend that all contributions in a volume be written in the same LaTeX version, preferably LaTeX2e. To ensure best results, authors are asked to use the LaTeX2e style files available from Springer's web-server at

   ftp://ftp.springer.de/pub/tex/latex/svmonot1/ (for monographs) and
   ftp://ftp.springer.de/pub/tex/latex/svmultt1/ (for summer schools/tutorials).

   Additional technical instructions are available on request from: lnm@springer.com.
5. Careful preparation of the manuscripts will help keep production time short besides ensuring satisfactory appearance of the finished book in print and online. After acceptance of the manuscript authors will be asked to prepare the final LaTeX source files and also the corresponding dvi-, pdf- or zipped ps-file. The LaTeX source files are essential for producing the full-text online version of the book. For the existing online volumes of LNM see: http://www.springerlink.com/openurl.asp?genre=journal&issn=0075-8434.

   The actual production of a Lecture Notes volume takes approximately 12 weeks.
6. Volume editors receive a total of 50 free copies of their volume to be shared with the authors, but no royalties. They and the authors are entitled to a discount of 33.3% on the price of Springer books purchased for their personal use, if ordering directly from Springer.
7. Commitment to publish is made by letter of intent rather than by signing a formal contract. Springer-Verlag secures the copyright for each volume. Authors are free to reuse material contained in their LNM volumes in later publications: a brief written (or e-mail) request for formal permission is sufficient.

**Addresses:**

Professor J.-M. Morel, CMLA,
École Normale Supérieure de Cachan,
61 Avenue du Président Wilson,
94235 Cachan Cedex, France
E-mail: Jean-Michel.Morel@cmla.ens-cachan.fr

Professor F. Takens, Mathematisch Instituut,
Rijksuniversiteit Groningen, Postbus 800,
9700 AV Groningen, The Netherlands
E-mail: F.Takens@rug.nl

Professor B. Teissier,
Institut Mathématique de Jussieu,
UMR 7586 du CNRS,
Équipe "Géométrie et Dynamique",
175 rue du Chevaleret,
75013 Paris, France
E-mail: teissier@math.jussieu.fr

*For the "Mathematical Biosciences Subseries" of LNM:*

Professor P.K. Maini, Center for Mathematical Biology,
Mathematical Institute, 24-29 St Giles,
Oxford OX1 3LP, UK
E-mail: maini@maths.ox.ac.uk

Springer, Mathematics Editorial I, Tiergartenstr. 17,
69121 Heidelberg, Germany,
Tel.: +49 (6221) 487-8259
Fax: +49 (6221) 4876-8259
E-mail: lnm@springer.com